CLINICAL USES OF FLUORIDES

*A State of the Art Conference
on the Uses of Fluorides
in Clinical Dentistry*

CLINICAL USES

May 11 and 12, 1984
Holiday Inn, Union Square
San Francisco, California

Conference Director
Stephen H. Y. Wei, D.D.S., M.D.S.

Sponsored by

Department of Growth and Development
School of Dentistry
University of California, San Francisco

The United States Public Health Service, Region IX

Council on Dental Research and Council on Dental Therapeutics of
the American Dental Association
in cooperation with Postgraduate Dentistry
School of Dentistry
University of California, San Francisco

OF FLUORIDES

A State of the Art Conference on the Uses of
Fluorides in Clinical Dentistry

Edited by

Stephen H. Y. Wei, D.D.S., M.S., M.D.S., F.I.C.D., F.R.A.C.D.S.

Professor and Head
Department of Children's Dentistry and Orthodontics
University of Hong Kong

Lea & Febiger

1985 Philadelphia

Lea & Febiger
600 Washington Square
Philadelphia, PA 19106-4198
(215) 922-1330

Library of Congress Cataloging in Publication Data
Main entry under title:

Clinical uses of fluorides.

"Sponsored by Department of Growth and Development, School of Dentistry, University of California, San Francisco, the United States Public Health Service, Region IX [and] Council on Dental Research and Council on Dental Therapeutics of the American Dental Association in cooperation with Postgraduate Dentistry, School of Dentistry, University of California, San Francisco."
   1. Dental caries—Prevention—Congresses. 2. Fluorides—Therapeutic use—Congresses. I. Wei, Stephen H.Y. II. University of California, San Francisco, Dept. of Growth and Development. III. United States Public Health Service. Region IX. IV. Council on Dental Research (U.S.) [DNLM: 1. Fluorides, Topical—therapeutic use—congresses. 2. Fluorides—therapeutic use—congresses. QV 50 S797c 1984]
RK331.C55 1984   617.6'01   84-19374
ISBN 0-8121-0970-8

Although the author and publisher have taken every precaution to assure that the dosages are accurate, they assume no responsibility for the use of these drugs or dosages under any circumstances. The reader is advised to consult the drug package insert for current accepted indications and methods of drug use before using any drug.

PRINTED IN THE UNITED STATES OF AMERICA

Print No. 4 3 2 1

PREFACE

A plethora of new fluoride products are now available. Dentists and dental hygienists are increasingly unsure of the relative efficacy of each new or combination agent introduced for professional and/or home use. There is controversy among researchers and clinicians concerning the claims of these new products, particularly in the areas of patient acceptance, modes of administration, mechanisms of action, and clinical effectiveness. Furthermore, the role of fluorides may be increasingly important in remineralization of incipient caries, root surface caries, dentinal hypersensitivity, prevention and treatment of gingivitis, and certain forms of periodontal disease.

The purpose of this book is to update and critically review past and current fluoride agents and methods of application. The authors have attempted to clarify the existing confusion and to propose resolutions to key controversies in all areas of fluoride use by dental personnel. The newest agents, modalities of delivery, and mechanisms of actions of fluorides are discussed in the context of clinical effectiveness and FDA approval or censure. In addition, the authors have put the latest research findings in clinical perspective.

The conference and this publication were made possible by educational grants to the Division of Pediatric Dentistry, Department of Growth and Development, School of Dentistry, University of California, San Francisco by the dental industry. The conference was jointly sponsored by the U.S.P.H.S. Regional Office, Region IX and the Council on Dental Therapeutics and the Council on Dental Research of the American Dental Association. Special invited guests from the California Dental Association, U.S.P.H.S., the American Academy of Pedodontics and the American Society of Dentistry for Children were in attendance to help disseminate the information.

Stephen H. Y. Wei, D.D.S.
Hong Kong

v

ACKNOWLEDGEMENTS

This conference and publication were made possible by educational grants to the University of California, San Francisco by:

Block Drug Company, Inc.
John O. Butler Company
Colgate-Palmolive Company
Hoyt Laboratories
Johnson & Johnson Health Care
Mead Johnson & Company
Oral B
Procter & Gamble Company
Scherer Laboratories, Inc.
Warner-Lambert Company

CONTRIBUTORS

J. Michael Allen, D.D.S.
Captain, Dental Corps
United States Navy
Washington, D.C.

James W. Bawden, D.D.S., Ph.D.
Alumni Distinguished Professor
School of Dentistry
University of North Carolina
Chapel Hill, North Carolina

Harry M. Bohannan, D.M.D.
Dalton McMichael Fellow and Research Professor
School of Dentistry
University of North Carolina
Chapel Hill, North Carolina

Robert Boyd, D.D.S.
Assistant Professor, Department of Growth and Development
University of California, San Francisco
San Francisco, California

James P. Carlos, D.D.S.
Director, Epidemiology and Oral Disease Prevention Program
National Institute for Dental Research
Bethesda, Maryland

Fermin A. Carranza, Jr., Dr. Odont.
Professor and Chairman, Department of Periodontics
School of Dentistry
University of California, Los Angeles
Los Angeles, California

Rella R. Christensen, R.D.H.
   Co-Director, Clinical Research Associates
   Provo, Utah

James J. Crall, D.D.S.
   Assistant Professor, Department of Pedodontics
   College of Dentistry
   The University of Iowa
   Iowa City, Iowa

John C. Greene, D.D.S.
   Dean, School of Dentistry
   University of California, San Francisco
   San Francisco, California

Jon T. Kapala, D.M.D.
   Professor and Chairman, Department of Pediatric Dentistry
   Boston Graduate School of Dental Medicine
   Boston, Massachusetts

Katherine Kula, D.D.S.
   Assistant Professor, Department of Pediatric Dentistry
   Dental School
   University of Maryland at Baltimore
   Baltimore, Maryland

Corrine H. Lee, R.D.H.
   Dental Hygienist, Honolulu Head Start
   Honolulu, Hawaii

Weyland Lum, D.D.S.
   Private Practice in Pedodontics
   San Francisco, California

Martin L. MacIntyre, D.D.S.
   Regional Dental Consultant, Region IX
   U.S. Public Health Service
   San Francisco, California

John E. Mazza, D.D.S.
   Lecturer, Department of Periodontics
   School of Dentistry
   University of California, Los Angeles
   Los Angeles, California

Robert Mecklenburg, D.D.S.
   Chief Dental Officer, U.S.P.H.S.
   Rockville, Maryland

Edgar W. Mitchell, Ph.D.
Secretary, Council on Dental Therapeutics
American Dental Association
Chicago, Illinois

Conrad A. Naleway, Ph.D.
Research Chemist, American Dental Association
Chicago, Illinois

Ernest Newbrun, D.M.D., Ph.D.
Professor, Division of Oral Biology
Department of Stomatology
University of California, San Francisco
San Francisco, California

Michael G. Newman, D.D.S.
Adjunct Professor, Department of Periodontics
School of Dentistry
University of California, Los Angeles
Los Angeles, California

Arthur J. Nowak, D.M.D.
Professor, Department of Pedodontics
College of Dentistry
University of Iowa
Iowa City, Iowa

Dorothy A. Perry, R.D.H., Ph.D.
Adjunct Assistant Professor, Department of Periodontics
School of Dentistry
University of California, Los Angeles
Los Angeles, California

Louis W. Ripa, D.D.S.
Professor and Chairman, Department of Children's Dentistry
State University of New York at Stonybrook
School of Dental Medicine
Stonybrook, New York

J. Keith Roberts, D.D.S.
Private Practice in Pediatric Dentistry
Bloomington, Indiana

Michael Roberts, D.D.S.
Chief, Patient Care Section, CIPCB
National Institute of Dental Research
Bethesda, Maryland

Leon M. Silverstone, D.D.Sc., Ph.D.
    Associate Dean for Research
    University of Colorado Health Sciences Center
    Denver, Colorado

Neil Smithwick, D.D.S.
    13th District Trustee
    American Dental Association
    Sunnyvale, California

William R. Snaer, D.D.S.
    Private Practice in Pediatric Dentistry
    Pasadena-Arcadia, California

John W. Stamm, D.D.S.
    Professor and Chairman, Department of Community Dentistry
    McGill University
    Quebec, Canada

George K. Stookey, Ph.D.
    Professor, Department of Preventive Dentistry
    Indiana University School of Dentistry
    Indianapolis, Indiana

Norman Tinanoff, D.D.S.
    Associate Professor, Department of Pediatric Dentistry
    The University of Connecticut Health Center
    Farmington, Connecticut

Carl A. Verrusio, Ph.D.
    Secretary, Council on Dental Research
    American Dental Association
    Chicago, Illinois

James S. Wefel, Ph.D.
    Associate Professor, Department of Pedodontics and Dow's Research Institute
    The University of Iowa
    Iowa City, Iowa

Stephen H. Y. Wei, D.D.S.
    Professor and Head, Department of Children's Dentistry and Orthodontics
    The Prince Philip Dental Hospital
    University of Hong Kong
    Hong Kong
    Formerly Professor and Chairman, Division of Pediatric Dentistry
    Department of Growth and Development, School of Dentistry
    University of California, San Francisco
    San Francisco, California

Stephen S. Yuen, D.D.S.
    Private Practice
    Hayward, California

INVITED GUESTS

Robert W. Beck, D.D.S.
Chief, Dental Head Start Program, Rockville, Maryland

William B. Bock, D.D.S.
Chief, Dental Disease Prevention Activity, Centers for Disease Control, Atlanta, Georgia

James Clark, D.D.S.
President, American Society of Dentistry for Children, Dubuque, Iowa

Richard Ahlfeld, D.D.S.
School of Dentistry, University of The Pacific, San Francisco, California

Alice Horowitz, R.D.H.
National Institute of Dental Research, Bethesda, Maryland

Herschel S. Horowitz, D.D.S.
Chief, Clinical Trials Section, National Institute of Dental Research, Bethesda, Maryland

Robert Isaacson, D.D.S., Ph.D.
Professor and Chairman, Department of Growth & Development, School of Dentistry, UCSF, San Francisco, California

Reginald Louie, D.D.S.
Regional Dental Consultant, Region IX, National Health Service Corps, San Francisco, California

Dale F. Redig, D.D.S.
Executive Director, California Dental Association, Sacramento, California

Paul Robertson, D.D.S.
     Professor and Chairman, Department of Stomatology, University of California, San Francisco, San Francisco, California

Ira Shannon, D.D.S.
     Professor of Biochemistry, University of Texas, and Director, Oral Disease Research Lab, VA Medical Center, Houston, Texas

Samuel J. Wycoff, D.D.S.
     Department of Community Dentistry, University of California, San Francisco, San Francisco, California

CONTENTS

SECTION V

PANEL DISCUSSIONS, 203

Robert Boyd
Rella R. Christensen
Jon T. Kapala
Corrine H. Lee
Weyland Lum
Martin L. MacIntyre
Michael Roberts
J. Keith Roberts
William R. Snaer
Stephen S. Yuen

Chair

Stephen H.Y. Wei

OPEN FORUM FOR QUESTIONS AND DISCUSSIONS, 217

INDEX, 227

Introduction of Guests, Speakers, and Purpose of Conference

INTRODUCTION AND PURPOSE OF CONFERENCE

Stephen H.Y. Wei

On behalf of the planning committee, the School of Dentistry, University of California, San Francisco, and the American Dental Association (ADA), I want to welcome all of you to sunny California and the wonderful city of San Francisco. We are all looking forward to two days of listening to rigorous scientific presentations and stimulating discussions, resolving some controversies, and learning how to utilize this new information in our clinical practices.

It has been 10 years since the last major conference on fluorides was held at the International Workshop on Fluorides and Dental Caries Prevention in Maryland. Since then, many new agents have been introduced to the dental profession (e.g., Fluorigard, ACT, and the commercially available 0.4% SnF_2 gel). The antiplaque aspect of fluorides was not fully explored nor was it commonly believed that fluorides had any impact on gingivitis or periodontal disease. The importance of fluorides in remineralization is certainly more recognized today than it was 10 years ago. Many dentifrice manufacturers now claim that one of the major mechanisms of action of their product is the ability of their dentifrice formulation to remineralize incipient enamel caries. At the last conference, the time-release fluoride system developed by the National Institute for Dental Research was not even in a conceptual stage, and fluoride-impregnated dental floss and resins were in the developmental stages. There has been considerable research and development since the last conference. It is, therefore, timely and necessary that a comprehensive review of all this progress be made now.

The emphasis of this conference is on the clinical uses of fluorides, particularly as they apply to the dental office. Dental hygienists and dentists are often confused and unable to make individual judgements on which products are better and which are more cost effective. They find that there are conflicting claims of one product over another. The dental profession has to select between flavor, viscosity, cost, patient acceptance, and claims of laboratory superiority vs. clinical effectiveness. It is the aim of this conference to address many of these practical questions.

The necessity of organizing this conference was justified by the overwhelming response of the number of registrants. We had planned for a total attendance of approximately 200 registrants. In the end, we had to limit the attendance to 300, although we could have accepted more than 350. The number of registrants shows the high intensity of interest in this topic.

When I contacted the speakers, they accepted their assignments with enthusiasm. They agreed with me that the topics selected for them needed intensive review and clarification. The speakers have all done an outstanding job, as witnessed by the superior quality of their manuscripts. We are indeed fortunate to have outstanding moderators, including Drs. Jim Bawden, Harry Bohannan, Carl Verrusio, and Captain Mike Allen. All of the moderators have considerable interest in this topic, since they have been or are currently responsible for preventive programs as clinicians, program administrators, and researchers in this field.

Industry has a strong interest in these topics as well. Their generous support of the conference is deeply appreciated and I'm pleased to recognize representatives from each of the ten companies present. Our thanks goes to Dr. Richard Brogle of the Block Drug Company Inc., Mr. Bud Tarrson of the John O. Butler Company, Dr. Tony Volpe from the Colgate-Palmolive Company, Mr. Hod Moses from Hoyt Laboratories, Dr. Hazen Baron from the Johnson & Johnson Health Care Company, Dr. George Baker from Mead Johnson & Company, Mr. Bob Perry from the Oral B Company, Mr. Bob Lehnhoff from the Procter & Gamble Company, Mr. Dennis Groat from Scherer Laboratories, and Mr. Joe Clark from the Warner Lambert Company.

I'd like to thank the members of the planning committee, which is composed of Dr. Marty MacIntyre of the U.S. Public Health Service, Dr. Weylund Lum, a pediatric dentist from San Francisco, Dr. Edgar Mitchell of the Council on Dental Therapeutics, and Dr. Carl Verrusio from the Council on Dental Research of the ADA.

Finally, I'd like to thank Ms. Betty Rojas and her staff from Postgraduate Dentistry, UCSF, for assisting us in organizing this conference, and last but not least, to Mrs. Martie Van Gorp, my secretary, for the tremendous amount of hard work that she has done in numerous ways for this conference.

It is now my pleasure to introduce Dr. John Greene to extend the first official welcome from the University of California, San Francisco.

INTRODUCTION AND WELCOME

John C. Greene

We at the University of California, San Francisco are pleased to collaborate with the ADA and the United States Public Health Service in sponsoring this Conference on Clinical Uses of Fluorides.

In recent years, dramatic progress has been made in the prevention of dental caries in this country and abroad. Reports from around the world attest to the rapid decline in caries prevalence, especially in industrialized nations. But, while we are congratulating ourselves for our victories, it would be tempting to become complacent and assume that the caries battle is over. We should, however, realize that the reductions we are now seeing will not automatically continue on a straight line until caries disappears. Sustained efforts will be needed by all to keep the encouraging trends moving in the right direction. Complacency must be avoided until this disease is eliminated as a significant health problem. I am sure that this audience would agree that carious lesions are still occurring too frequently, and that as long as they are, we must continue our efforts at prevention.

The purpose of this conference is to facilitate the achievement of maximum benefit from clinical uses of our most effective caries prevention agent—FLUORIDE. Additionally, the effect of fluoride on gingivitis and periodontal health will be explored. This area has received too little attention until recently.

If this conference is successful in achieving its purpose, we will leave here with a better understanding of the uses of fluorides in clinical dentistry, and some of the confusion surrounding this area will have been clarified.

I want to congratulate and express my appreciation to Dr. Wei and the others for organizing this conference, and for assembling such an impressive array of experts to discuss this most timely subject.

WELCOME ON BEHALF OF THE AMERICAN DENTAL ASSOCIATION

Neil Smithwick

The ADA is pleased to join with our colleagues of the University of California and the United States Public Health Service in sponsoring this landmark conference on the uses of fluorides in clinical dentistry. Chairman Wei has brought together premier researchers and experts in the field, making this conference truly "State of the Art." I bring you the personal greetings of President Don Bentley and the Board of Trustees, and we wish you a most productive, informative, and stimulating conference.

WELCOME FROM UNITED STATES PUBLIC HEALTH SERVICE

Robert Mecklenburg

It is a pleasure to be with you and to extend greetings from Dr. Edward Brandt, the Assistant Secretary for Health, and Dr. C. Everett Koop, the Surgeon General of the United States Public Health Service. The Public Health Service is privileged to be a co-sponsor for this important meeting. This meeting is another example of the spirit of cooperation that exists in dental affairs between the federal government, dental education, industry, and the practicing profession. Such cooperation has in the past served to effectively protect and improve the health of the public.

It is now obvious to everyone that the oral health of the American people has substantially improved during the latter part of this century. This has not happened by accident, but through the determination of many health workers and the support of a concerned public.

Concern for oral health rose to national importance during the late 1940s. Before then, dental problems had been understood to be primarily a personal matter. Then, during the Second World War, it was learned that the single greatest reason for rejecting people from military duty was poor oral health. Within the services, severely diseased teeth and periodontal tissues disabled an unacceptably high number of personnel during maneuvers and while in action. Oral health became a national security issue.

Population surveys were conducted after the war that showed that over 95% of people experienced dental caries. By ages 20 to 24 the average person had already lost 4 teeth. Half of the population had periodontal disease by age 50, and half were edentulous in at least one arch.

Once oral health became recognized as a public health problem, an aroused public and a concerned profession took their case to the United States Congress. The congress, and people across the land, heard that dental caries was so severe that, "If all of the dentists in the nation were to restore decayed teeth beginning on the east coast, and work their way west, by the time they reached Pittsburgh, Pennsylvania, they would have to return to the east coast and begin again." As a consequence, the congress established federal organizations within the Public Health Service that would help states, communities, and the practicing profession to resolve the problem.

In June, 1948, the United States Public Health Service, National Institute of Dental

7

Research was established as the third institute within the National Institutes of Health. One month later, in July, 1948, one million dollars were funded for dental activities—a considerable sum in those days. A year later, in June, 1949, a division of Dental Public Health and a division of Dental Resources were established. In June, 1950, the surgeon general released a declaration in support of water fluoridation as being the single most effective, safe, and least costly means of preventing dental caries.

We all know what happened then. Public health dentists, community health workers, and community leaders began the long process of building support for fluoridation. In time, over half of the population received the benefits of drinking fluoridated water. During these same years, research produced a variety of alternative means to deliver fluoride ion where needed. Today, we can review such history with a great sense of accomplishment. The people have been served. Dental caries prevention measures are estimated to be saving from 4 to 5 billion dollars worth of treatment expenses each year, not counting considerable intangible public benefit.

Then why are we here? Because we recognize that the dental profession will not leave a work only half done. It is technically possible for dental caries to be a rare disease using preventive measures already available. In recent years, the prevalence of dental caries in 17-year-olds declined from 17 to 11 teeth—A very satisfying trend. The dental profession and the public, however, can be no more content with the level achieved than would be the medical profession, if, for example, a chronic, progressive bone disease gradually destroyed 11 rather than 17 fingers and toes.

This conference will provide a variety of methodologies and the scientific bases for them so that oral health status gains can be protected and further progress achieved. A growing body of scientific evidence indicates that a lifetime exposure to minute levels of fluoride is necessary to maintain protection of the teeth against dental caries. It is the maintenance of a consistent level of optimum exposure that must hold the center of our attention, and not the means used to attain a level. It remains the art of the profession to orchestrate the combination of methods so that their uses are most acceptable, feasible, and accessible to the public. No single method may be best under all conditions.

You are to be congratulated on your discontent with the level of oral health already attained, and for your desire to share information and seek out truth through scientific inquiry. Your presence is witness to your spirit of inquiry and sense of professional responsibility.

On behalf of the Public Health Service, I commend Dr. Wei, for assembling so many distinguished speakers; the University of California at San Francisco and the industry for their interest and support, and all those who have worked so diligently to make this conference possible.

Section I

JAMES BAWDEN, MODERATOR

Chapter 1

AMERICAN DENTAL ASSOCIATION'S PROGRAM OF ACCEPTANCE OF FLUORIDE PRODUCTS

Edgar W. Mitchell

In 1930, the American Dental Association (ADA) established an agency to evaluate dental drug products and dental cosmetic agents. This agency, which later became the Council on Dental Therapeutics, was also directed to provide information about dental products to both the dentist and the public. Another function of this agency was to encourage research on the products when more information was needed. The Council was established because the Association believed there was a need for professional evaluation and commentary on the confusing array of "medicaments" and "patent medicines" that were being promoted and sold. Advertising claims for some of the products were misleading; some of the products were of questionable value, and some were dangerously toxic. Discussion will focus on whether such a program can be of as much or of more value to the dental profession today as it was in 1930. An overview of the current acceptance program for dental drugs, including fluoride products, will be presented.

The Council on Dental Therapeutics of the ADA was established in 1930 with a directive by the House of Delegates to study, evaluate, and disseminate information about dental therapeutic agents, their adjuncts, and dental cosmetic agents that are offered to the public or to the dental profession. In addition, the Council was to encourage research in the field of dental therapeutics in the academic and industrial research communities. At the present time the Council's activities are directed primarily through five programs in several areas. These programs are:

1. Health Assessment Program
2. Mercury Testing Service
3. Conducting of symposia and meetings sponsored or cosponsored by the Council in areas of drugs and other topics of relevant interest to the profession
4. Communication of the findings of the Council's activities
5. Product Evaluation and Acceptance Program

HEALTH ASSESSMENT PROGRAM

The Health Assessment Program has been conducted by the Council since the early 1960s. This program is conducted at each Annual Session. It is financially supported by the American Fund for Dental Health and corporate sponsors. It essentially pro-

11

vides a health evaluation for any dentist who is registered at the Annual Session of the ADA. One goal of this program is to raise the dentists' awareness of their own health and to encourage them to follow-up this screening with a complete examination by their personal physician. Through this program the ADA has learned the health and disease characteristics of dentists, including their use of personal medications. From serum samples collected through this program, the ADA learned that its members have a relatively high prevalence of periodontal disease and hepatitis B. This information led to a strong recommendation for the profession when hepatitis B vaccine became available. The Association also learned from the Health Assessment Program that dentists are as healthy as the general population with respect to cardiovascular disease.

MERCURY TESTING SERVICE

Another aspect of the Health Assessment Program was the urinary evaluation of mercury, which led to a second program conducted by the Council. This program is the Mercury Testing Service. It was clearly demonstrated from the mercury evaluation conducted at the Health Assessment Program that dentists excrete urinary levels of mercury 3 to 5 times higher than those of the general population, owing to their occupational exposure to mercury through the placement and removal of dental mercury amalgam.

The Council on Dental Therapeutics offers the Mercury Testing Service to dentists and their staff members on an enrollment/subscription basis. A fee has been established to recover the costs of conducting the program. Urine samples are received by mail and analyzed in our laboratory and results are sent to dentists and their staff members. When dentists enroll themselves or their staff members in this program, they receive materials for collecting urine samples and a questionnaire that determines basic practice characteristics of the dentists and the practice in which they are working. Six assays of urine are used to determine levels of mercury. Recommendations for incorporation of mercury hygiene procedures developed by the Council on Dental Materials, Instruments, and Equipment are discussed with each participant.

CONDUCTING SYMPOSIA AND MEETINGS

A third activity of the Council is the sponsorship of symposia, meetings, and workshops on relevant topics concerning therapeutics or practice characteristics that do not fall under the purview of any other agency of the Association. For example, in 1982 the Council, with the University of Illinois, sponsored a symposium on hepatitis B. That symposium resulted in the Council's recommendations for infection control guidelines for hepatitis and indicated the need for the use of the vaccine. Other symposia sponsored or cosponsored by the Council on Dental Therapeutics included a conference at the 1982 Annual Session on root surface caries: its prevalence, treatment, and control. A meeting was sponsored at the Annual Session in Dallas on patients with cardiovascular disease and how their dental treatment should be managed. A conference on clinical testing of anticaries agents was held at the ADA, in 1983. A symposium has been scheduled for the 1985 Annual Session in San Francisco on the topics of general anesthesia and conscious sedation in the ambulatory dental patient.

COMMUNICATIONS

A fourth area of activity of the Council is communicating with the profession and with the public. This communication is done primarily by five routes. Three of these involve publications of the Council and include *Accepted Dental Therapeutics,* which is now in its 39th edition. This book gives a listing of products and a brief description of the role of the products in dental practice. I will describe in more detail the section of that publication relating to fluoride products. The Council is revising this edition of the book and the 40th edition has been sent to the printer (1984 marks the 50th anniversary of this book). Other publications include reports in the *Journal of the American Dental Association,* which are published in the Council reports section. These reports become official recommendations and documents of the Association's councils. An example of a Council report is "Infection Control in the Dental Office," which was published in 1978 and has now been incorporated into *Accepted Dental Therapeutics,* which is published by the Council on Dental Therapeutics. Each year an updated listing of accepted products is published in the Journal. The third means by which the Council publishes many of its recommendations and guidelines is in the *ADA News,* which is published once every two weeks. In recent issues of the *ADA News* the Council has published statements to the profession on the status of the Nisentil brand of alphaprodine as an analgesic in the conscious sedation technique, and the status of FD&C Red #3 as a plaque disclosing agent.

The Council on Dental Therapeutics also responds to many media requests, from newspapers, monthly magazines, and news programs. The Council assists other staff members of the ADA in preparing responses to media inquiries. The Council also assists in the preparation of brochures sold to dentists for use in discussing various aspects of practice with their patients.

PRODUCT ACCEPTANCE PROGRAM

The fifth and most significant portion of the Council's activities in terms of time and effort is devoted to the Product Acceptance Program. The purpose of this program is to review and evaluate the safety and efficacy of products, and to relay the results of this evaluation to the dental profession and the public. The testing of products for these criteria is conducted by manufacturers, who submit their products on a voluntary basis for consideration in this program.

Some product testing is conducted in the laboratory of the Council's Division of Chemistry at the ADA. Dr. Naleway, who is Director of that Division, outlines in Chapter 10 the types of analytical studies that are conducted for the Council. These studies are generally requested by the Council when there is a question about the ability of a product to deliver the active ingredient or to verify some unique property of the product. A significant amount of the testing currently conducted in the Council's laboratory is in the area of fluoride products. At the present time, every fluoride product newly submitted to the Council and every one resubmitted for renewal of acceptance are tested for fluoride content. Some products are also examined for their ability to deliver fluoride to the tooth. Only fluoride tablets are not tested routinely. The Council has currently classified accepted fluoride products in six categories. These are:

    I.   Dietary Fluoride Supplements
           (1) tablets—9 products
           (2) liquids (drops or rinses)—8 products

II. Sodium Fluoride (NaF) Rinses
    (1) once a day use—0.05% NaF—8 products
    (2) weekly use—0.2% NaF—3 products

III. Acidulated Phosphate Fluoride (APF) Solutions
    (1) solutions for office paint on or tray use—1.23% F, 1% H_3PO_4—6 products
    (2) rinses for daily use—0.05% F, 0.01% H_3PO_4—2 products

IV. Acidulated Phosphate Fluoride Gels
    (1) office tray application use—1.23% F, 1% H_3PO_4—27 products

V. Stannous Fluoride (SnF_2) Preparations
    (1) SnF_2 rinse—daily use—1 product
    (2) SnF_2 gels—daily use—3 products

VI. Fluoride Dentifrices
    (1) NaF—2 products
    (2) Na_2PO_3F—5 products

Fluoride products in categories not currently accepted by the Council include stannous fluoride rinses for office use, acidulated phosphate fluoride rinse products for office use, and fluoride prophylaxis pastes. To date, products submitted to the Council in these categories have not been submitted with adequate clinical studies to support their efficacy. The Council, therefore, has had no basis upon which to evaluate or predict their clinical benefit.

Previously, all fluoride products evaluated by the Council were supported by clinical studies employing the actual products submitted. This is no longer true in several of these categories, and clinical studies may not be required in other categories in the future if the Council continues to recognize results from laboratory tests as adequate to predict clinical efficacy. Products now accepted by the Council based on previous studies with essentially equivalent formulations include the fluoride supplements, sodium fluoride rinses for daily and weekly use, some acidulated phosphate fluoride products, and stannous fluoride rinses and gels for daily use.

The Council is currently considering the acceptance of fluoride dentifrices based on laboratory data. Guidelines on how the Council plans to review dentifrices using laboratory methods have been circulated to interested parties for comment and will likely be adopted this year. The Council has now directed its staff to study the feasibility of evaluating fluoride prophylaxis pastes based on laboratory data. New products with new active fluoride ingredients, or existing products proposed for uses or in dosages other than those supported by existing clinical studies will still require clinical evaluation for review by the Council.

The Council has indicated that it is *also* within its charge to test and evaluate fluoride products which may not have been submitted by a manufacturer. This evaluation will be an extension of its new guidelines, and will essentially be a commentary on whether a product has adequate levels of bioavailable fluoride and whether that fluoride can be predicted to reach its target organ, the tooth.

An end result of a manufacturer's participation in the acceptance program is the ability to use the Council's Seal of Acceptance on products that are promoted and sold both to the public and to the profession. It is the *intent* of the Council and the

Association to indicate to dentists and the public that the products that bear the Seal of Acceptance have been shown to be safe and effective, and that information on labels and promotional material for the product is accurate. When products bearing the Seal are promoted in a questionable manner, the dental profession should challenge these claims by calling the ADA. As part of the acceptance program the Council and staff also review advertising for accepted products. The reader of that advertising, whether it appears in the public or dental media, may be assured that all claims in that advertising are accurate and not misleading. The Council hopes that you, as professional users of dental products, will look for the Seal of Acceptance on products you use in your practice, and that you will recommend products bearing the Seal of Acceptance to your patients. We want what you want—effective products that patients will use properly to achieve maximum results.

Chapter 2

CRITICAL ASSESSMENT OF PROFESSIONAL APPLICATION OF TOPICAL FLUORIDES

James S. Wefel

This paper reviews the techniques of application and use for sodium fluoride (NaF), stannous fluoride (SnF_2), and acidulated phosphate fluoride (APF) products. The properties of acidulated phosphate fluoride gels are discussed including pH, concentration, viscosity, flavoring, and sweeteners. Changes in original formulations, such as higher pH, lower concentration, high viscosity, and thixotropic characteristics, have been made. Implications on effectiveness of these new formulations are examined. The effectiveness of topical fluorides in adults, in fluoridated areas, for public health programs, and private practice are also covered.

Since the introduction of topically applied fluorides in the 40s, clinical testing of effectiveness, development of new agents, and development of various modes of application have been studied in the search for the optimal topical fluoride agent. Today, we may receive topical fluoride benefits from water fluoridation, fluoride supplements, fluoride dentifrices, fluoride rinses, and professionally applied agents in the dental clinic or office. Research literature contains hundreds of articles pertaining to the use of topical fluoride agents and the task of synthesizing that amount of knowledge is a tremendous one. Fortunately, this chapter is limited to the professional application of topical fluoride. Recent reviews of the techniques of application, products available, and clinical effectiveness have been carried out by Ripa,[1] Brudevold and Naujoks,[2] Clarkson and Wei,[3] Wei,[4] Horowitz,[5] and others. I will use these reviews to describe the known effectiveness and accepted methods of application for sodium fluoride, stannous fluoride, and acidulated phosphate fluoride, and try to add to this information in terms of products currently available, their compositional properties, and effectiveness in several clinical settings.

ACCEPTED TOPICAL FLUORIDE AGENTS

The 39th Edition of *Accepted Dental Therapeutics* contains a listing of the American Dental Association (ADA) accepted preparations for sodium fluoride solutions, stannous fluoride solutions, and APF solutions and gels. The initial formulation of topical fluoride agents was in the form of solutions, and therefore they had to be applied with a swabbing technique after thorough drying of the teeth. Differences

Table 2–1. Summary of Clinical Trials Employing Operator-Applied Aqueous Solutions of NaF, SnF$_2$ or APF to the Permanent Teeth of Children in Fluoride-Deficient Communities

Duration of Study (Year)	Number of Treatment Groups			Average Results % Caries Reduction (DMFS)		
	NaF	SnF$_2$	APF	NaF	SnF$_2$	APF
1	12	9	8	30	39	35
2	9	8	10	28	27	30
3+	4	1	9	29	8	20
1–3+	25	18	27	29	32	28

(From Ripa, L.W., Int. Dent. J., *31*:105–120, 1982.)

in fluoride concentrations, application techniques, length of application, frequency of treatment, and duration of study are several criteria that have varied in clinical trials and prevent the direct, critical comparisons of one study with another. Other important factors to be considered in the experimental design involve the population studied by such characteristics as age, socioeconomic status, caries incidence, and oral hygiene. The degree of variability of percentage caries inhibition reported in the literature may in part be due to these variables. One must still make some attempt to evaluate the overall effectiveness of these agents. The review article by Ripa[1] is perhaps the most recent and summarizes 35 clinical studies with 74 treatment groups. Operator-applied aqueous solutions of NaF, SnF$_2$, and APF were evaluated for clinical efficacy and resulted in similar findings. From Table 2–1 it can be seen that NaF gave a 29% reduction, SnF$_2$ a 32%, and APF a 28% reduction in DMFS.

Studies on the retained effectiveness of the fluoride agents are less numerous and have been summarized by Brudevold and Naujoks.[2] The results indicate that there is a decrease in effectiveness 2 years after treatment as may be expected, but more importantly, they show a significantly retained anticaries effect (10 to 40%) (Table 2–2). A longer lasting caries-reducing effect was noted for APF, and this was thought to be due to the greater amount of fluoride deposited in the enamel.

Another important aspect of topical fluoride applications is the effect on teeth erupting during the study. Although the numbers of patients or teeth upon which the

Table 2–2. Retained Anticaries Effect from Topical Fluoride Treatments

	Treatments		Length of Study (Years)	Years after Last Treatment	Subjects		Fewer DMFS, %	
	Agent	Number			Number	Age (Years)	Final	Post-Treatment
Bibby & Turesky, (1947)[6]	0.1% NaF	6	2	3	39	10 to 12	32	19
Syrrist & Karlsen, (1954)[7]	2% NaF	7	2	3	116	12	47	21
Sundvall-Hagland, (1955)[8]	2% NaF	4	1	2	107	3 to 5	19	12
Houwink et al., (1947)[9]	4% SnF$_2$	18	9	5	15	1 to 7	36	25
Horowitz & Kau, (1974)[10]	APF	3	3	2.5 to 3	108	10 to 12	35	31
	APF	6	—	2.5 to 3	92	10 to 12	49	43
	APF gel	3	—	2.5 to 3	105	10 to 13	26	21

(Modified from Brudevold, F., and Naujoks, R., Caries Res., *12*:52–64, 1978.)

Table 2–3   Effect of Topical Fluoride Treatment on Newly Erupted Teeth

	Treatment	Duration of Study (Years)	Caries Reduction	
			Base Line Teeth	Erupting Teeth
Averill et al., (1967)[11]	Topical NaF	2	12	43
Horowitz & Heifetz, (1969)[12]	Topical SnF_2	1	21	61
De Paola & Mellberg, (1973)[13]	APF prophylaxis paste	2	21	36
Downer et al., (1976)[14]	F prophylaxis paste + APF gel + Na_2PO_3F dentifrice	3	31	56

(From Brudevold, F., and Naujoks, R., Caries Res., *12*:52–64, 1978.)

results are made are much smaller, the caries-reducing effect has been shown to be greater in newly erupted than in mature teeth (Table 2–3). This effect is independent of the type of fluoride used, has been reported to occur with various fluoride vehicles, and is in the 40 to 50% range.[3,12]

Even though the clinical efficacy of these agents appears to be equal, APF products have become the agent of choice. This is due in large measure to the favorable characteristics of APF when compared to SnF_2, and the incorporation of gelling agents. Tray applications of APF gel have decreased the necessity for one-to-one applications and time required for dental personnel and patient during the topical fluoride treatment. Thus APF gel tray applications are convenient, save time, and appear to be in widespread use in the United States.

APF Gels

Compositional Changes.   The original APF solution contained 1.23% F from NaF and HF and 0.1 M orthophosphoric acid with a resulting pH of 3.2. This composition provides high concentrations of fluoride in an acidic environment to promote fluoride uptake. The phosphate was thought to provide a common ion effect and to favor fluorapatite formation. There are six accepted APF solutions listed in the 39th Edition of *Accepted Dental Therapeutics* and 27 APF gels. The gel formulations are to contain 1.23% F, 1% H_3PO_4, and a cellulose gelling base. The pH is stated to be 3.2 to 3.4, but the accepted levels are 3 to 4 with a viscosity of 7,000 to 20,000 centipoises.

Lower Concentrations and pH.   Shannon and Edmonds[15] used enamel solubility reductions to study the effects of diluting APF preparations fivefold. All diluted APF solutions were maintained at a pH of 3, and no loss of the solubility reducing effect of APF was observed with a concentration as low as 0.25% F. The 0.12% F solution showed considerably less protection in these studies. In a similar study, Shannon and Edmonds[16] also evaluated changes in pH and fluoride concentration. Several commercial oral rinses were tested, which had higher pH levels since they were designed for home use. The important point is that when the pH of 0.6% F solution was raised to 4.9 the enamel solubility reduction was greatly reduced.

Several studies in the Netherlands were performed with lower fluoride concentration solutions as well. Dijkman et al.[17] compared the fluoride uptake from APF solutions with fluoride concentrations of 1.23% to 0.11% and a pH of 4.0. No significant difference in the amount of fluoride acquired "on" and "in" the enamel was found between the different gel preparations. Lesion depth and mineral content of the lesion were the parameters investigated by Sluiter and Purdell-Lewis[18] when comparing APF applications containing 0.4%, 0.1%, and 0.01% F. These workers found that only the APF solution containing 0.4% F gave shallower lesions and concluded that 0.4% F should be the smallest fluoride concentration used for topical gel applications.

Since the in vitro testing just mentioned is not conclusive in terms of caries reduction, one must await clinical testing before a definitive statement may be made. At the recent IADR Meeting in Dallas, Hagan et al.[19] presented 2-year results on a half strength (0.6% F) and full strength (1.23% F) APF thixotropic gel. The study was conducted in a nonfluoridated community in which the placebo group experienced a mean DMFS increment of 4.39 compared to those receiving APF: 3.08 for the 1.23% F group and 3.31 for the 0.6% F group. This was a statistically significant reduction of 30% and 25% compared to the control group, although the reductions were not statistically different from one another. The authors suggest that a larger sample size should be used to determine if the trend toward less effectiveness with the half-strength fluoride gel is a consistent finding.

Looking at both the in vitro and in vivo data, it appears to me that a half-strength topical fluoride agent may be as effective as the traditional 1.23% APF gel as long as the pH remains at 4.0 or below.

Viscosity. Most topical fluoride gels approved by the ADA use carboxymethyl cellulose as the gelling base. As mentioned earlier, the viscosity should fall in the wide range of 7,000 to 20,000 centipoises. Recently, thixotropic gels which have a variable viscosity with inconstant pressure have been introduced. The advantage of these agents is that under biting forces the gel should thin out and more easily penetrate between teeth. Also, when not under stress, the gel is more viscous and should remain in situ in the tray and not run down the patient's throat. In vitro testing of thixotropic and regular APF gels have shown similar fluoride uptake and retention.[20] Similar findings were reported by Wei and Connor[21] following in vivo applications, although both the conventional and thixotropic gels showed a return to pretreatment levels after 7 days. The clinical effectiveness of a thixotropic gel and an APF solution were compared by Cobb et al.[22] in a fluoride-deficient community with a high caries incidence. Unfortunately, both agents were applied with a cotton swab, to limit the variables of treatment. Despite the disadvantage of not applying the thixotropic gel under a biting force, both types of gel had a 35% decrease in caries incidence after 2 years. As mentioned earlier, Hagan et al.[19] noted a 30% and 25% reduction using full- and half-strength thixotropic APF gels.

These studies appear to show that thixotropic APF gels are equal to conventional gels, but not superior in either in vitro or in vivo testing. A question still remains regarding the ability of a thixotropic agent to penetrate interproximally. The practitioner may choose the type of APF gel based on cost, patient acceptance, and ease of application.

Multiple Applications. Another approach to increase the fluoride content in enamel and hopefully reduce caries has been multiple short-term applications. The attractiveness of this sort of regimen is that short-term treatment may provide lasting protective effects. Five consecutive daily treatments of an APF gel or an amine fluoride gel proved to be ineffective in reducing decay after 2 years. The amine fluoride gel, but not APF, reduced occlusal decay significantly.[23] Shern et al. stated that, "Although APF gel appears to be of little or no value in populations experiencing a low incidence of decay, it has been demonstrated to give substantial protection in populations experiencing high increments of caries." This statement may be applicable to clinical trials in the future.

Another study[24] compared 5-min APF gel tray applications on 10 consecutive school days with a daily rinsing in school and with the combined regimens. After 12 months, only the results from the combined treatment were statistically significant, although results from all three regimens were significant after 23 months. The study was conducted in a high caries area, as seen by the placebo group, which had a 9.11 ± 0.63 mean net DF surface increment. The DF surface increment was 7.37 ± 0.52 for the gel group after 2 years. The unweighted means gave savings of 1.29 surfaces for the gel above and 2.81 surfaces for gel and rinse. This was ~16% for gel and 34% for both. Due to the high caries incidence, a significant number of surfaces were saved from restoration.

In general the multiple applications of APF gel over short time periods has not been tremendously successful in the United States. Horowitz et al.[25] and Heifetz et al.[26] have reported on similar trials with self-administered prophylaxis and toothbrushing with APF in New Orleans and São Paulo. Fifteen brushings were carried out 5 times during each school year at approximately 2-month intervals. The APF solution was half-strength, while the APF gel was full-strength. The results showed no differences in the 2-year incremental caries scores in the New Orleans trial, and a significant 19 to 33% caries reduction in the high caries incidence area of São Paulo. As stated by Horowitz,[25] ". . . the conclusion is that supervised toothbrushing with fluoride at a frequency of approximately 5 times/year confers limited inhibition of dental caries." Likewise, Mellberg et al.[27] stated, "The results suggest that in a fluoridated area, *when caries activity is low,* it would be difficult to show a long-term anticaries effect by increasing the fluoride concentration in sound enamel by a short series of self-applied topical fluoride applications."

EFFECTIVENESS OF INDIVIDUALIZED PREVENTIVE THERAPY

How can we best use the information available on operator-applied topical fluorides? What is the effectiveness of these agents in light of the changing patterns of caries prevalence?

In general, more recent clinical trials (Table 2–4) have shown average caries reductions from operator-applied topical fluoride gels to be less than 30%. Indeed,

Table 2–4. Studies on Effectiveness of Operator-Applied Topical Fluoride Gels Since 1975 Showing Marginal Caries Reduction

Study	Age	N	Appl/Yr	Duration	% Reduction (DMFT)	% Reduction (DMFS)
Shern et al., (1976)[23]	6–13	468	5*	2	+0.5	—
Zahran, (1976)[28]	7–9	1027	2	4	3	4.5
Mainwaring and Naylor, (1978)[29]	11–15	1718	2	3	—	14
DePaola et al., (1980)[24]	12–14	128	10†	2	—	14
Haupt et al., (1983)[30]	9–13	1519	2	2	—	11

*5 Treatments applied on consecutive days in first year only.
†10 Treatments applied on consecutive days in first year only.

Ripa[1] summarized APF gel trials and noted that the caries reductions are "somewhat less (22%) than those reported from an APF aqueous solution paint-on technique." Whether this is due to the application technique, the gels themselves, or the fact that only one application a year was performed is not known. Clarkson and Wei[3] also reported an average caries reduction of 25% for APF gels. A study[30] designed to look at the effects of professional prophylaxis prior to topical fluoride therapy showed little or no benefit due to prior toothcleaning or topical fluoride application. It must be pointed out, however, that the control group in the above study[30] was not derived by a random sampling process and may have influenced the findings. It was suggested by Haupt et al. "that the treatment regimen would be more effective if it was targeted to caries susceptible individuals rather than to all children."

Populations With Low Caries Activity

What about situations in which caries activity is low? Two populations that may be looked at in this regard are adults and lifelong residents of fluoridated communities. In both cases, the number of clinical trials is limited and, therefore, conclusions may be tentative. Of the nine studies[31–39] using adults summarized by Swango[40] and previously summarized by Ripa,[1] five studies show less than a 16% reduction, while four revealed significant reductions. Since the effectiveness of topical fluorides in adults is equivocal, only adults who continue to have moderate to high caries rates should receive topical fluoride application. As stated by Clarkson and Wei,[3] "Those who are not prone to dental caries would benefit so little from the treatments that the time might be better spent reinforcing the other aspects of preventive dentistry." What is not known, however, is the effect of topical fluorides on root caries. If root caries would respond to topical fluoride therapy in a manner similar to that of coronal caries, then a case may be established for using operator-applied topical fluorides in this segment of the population as well. Clearly, more work needs to be done in this area.

A similar situation exists for the use of topical fluorides in a fluoridated community. Because of the low cost-benefit ratio, topical fluoride treatment cannot be recommended as a public health measure and should be applied only to those patients with active decay. Since the caries rate in fluoridated communities is already low (e.g., 1 DMFS/yr), the same clinical effectiveness, for example, of 30% for a professionally applied topical fluoride would save only 0.3% of a surface. However, clinical trials of the effectiveness of operator-applied topical fluorides in fluoridated areas show an average caries reduction of only about 12%.[1,4] Even if this value were statistically significant, it is not necessarily clinically significant. Thus, for lifelong residents of optimally fluoridated communities, the need for professionally applied topical fluoride should again be individualized. It may also be argued that, if the caries prevalence has decreased to the point where nonfluoridated communities appear as fluoridated ones in terms of DMFS, little clinical significance may result from the use of professionally applied topical fluorides in these communities as well. This is due in large measure to the low caries rate and the widespread use of various topical fluoride vehicles (rinses, dentifrices).

This bleak picture of clinically significant reductions in dental decay from the use of professionally applied topical fluorides needs to be tempered by consideration of some of the less tangible aspects of preventive fluoride therapy. Even though use of topical fluorides in low-caries-rate individuals may result in savings of only tenths of a tooth surface, how do we put a price or a value on an unrestored tooth? Once restoration begins, some sound tissue is normally removed and if leakage and recurrent caries occur, additional time, money, and possibly pain will be associated with that tooth. It is understandable that public health programs cannot endorse the use of professionally applied topical fluorides because of the significant cost and marginal caries reduction. The use of self-administered fluoride therapy appears to be more practical in these programs.

Another consideration of preventive therapy is the mode of action of topically applied fluoride. Assuming that one of the main mechanisms of action of fluoride is on the developing white spot lesion, then fluoride will be extremely helpful in preventing lesion progression. This may occur from the promotion of remineralization or decrease in demineralization, which results in lesion arrest or repair. In vitro experiments by Silverstone et al.[41] have shown that as little as 1 ppm F combined with calcium and phosphate will remineralize white spot lesions. This process of remineralization will counter the episodes of demineralization that occur daily. It may then be possible to prevent the histological white spot lesion from progressing to the point where it becomes clinically detectable and then to the point where it must be repaired. Assuming that most interproximal areas contain clinically undetected white spot lesions; it would then seem prudent to continue the application of topical fluorides to maximize the potential for remineralization. The only question in my mind is what is the most appropriate vehicle? Since episodes of demineralization are a daily occurrence and the concentration of fluoride required is quite small, it appears that high-frequency, low-concentration fluoride agents may be most effective. Inherent in this approach is the fact that the patient can be motivated to comply with the preventive regimen.

Populations With High Caries Activity

What procedures should be followed for the caries-active individuals? Some individuals, whether they are adults, lifelong residents of fluoridated areas, or special at-risk populations such as those with reduced salivary flow, need the benefit of preventive fluoride therapy. This therapy should be individualized for the patients' needs, ability to comply, and caries activity in relationship to age. I feel that fluoride therapy should be targeted to the high-risk patients, high-risk teeth, and high-risk surfaces of the individual. In the context of professionally applied topical fluorides in the office or clinic situation, topical fluorides should be applied whenever teeth are erupting into the mouth. The literature clearly shows that the greatest degree of protection, almost 50%, occurs with teeth erupting during the clinical trial.

Studies also show that fluorides benefit the smooth surfaces to a greater degree than surfaces with pits and fissures. This has a dual benefit in that not only is there less decay, but the surfaces saved are those most difficult to restore. The interproximal surfaces with their contacts and protected areas for bacterial colonization are high-risk surfaces. In this regard, tray applications of APF gel certainly contact the buccal and lingual surfaces, but may not cover the interproximal sites. An in vitro study in our laboratory[42] using a model system showed a definite area below the contact point that was not covered by either a thixotropic or conventional gel. There was no salivary dilution in the model system and, therefore, little gel coverage in this most susceptible site. I think we must return to the procedure of flossing the teeth, not only prior to the gel application, but during or after it. This will carry the gel into the interproximal site and not rely on salivary dilution to spread the application around. In order to be beneficial, fluoride must first get to the at-risk site to interact. Special techniques or modes of application may need to be developed for at-risk, medically compromised populations (see Chapter 13).

I would be remiss when describing preventive therapy by professionals if I did not mention the application of sealants. These effective agents should be used in a combined preventive therapy program. As I have already stated, fluoride is least effective on the pit and fissure area and, therefore, the use of sealants becomes an increasingly important aspect of preventive dentistry. Thus, for caries-active individuals, the use of APF gels, interproximal flossing with gel, and sealants are suggested modes of professionally applied preventive therapy.

ACKNOWLEDGEMENT

This investigation was supported in part by USPHS grant DE04486 from the National Institute of Dental Research, National Institutes of Health, Bethesda, MD 20205.

REFERENCES

1. Ripa, L.W.: Professionally (operator) applied topical fluoride therapy: A critique. Clin. Prev. Dent., *4*:3–10, 1982.
2. Brudevold, F., and Naujoks, R.: Caries-preventive fluoride treatment of the individual. Progress in caries prevention. Caries Res., *12*(Suppl. 1):52, 1978.
3. Clarkson, B.H., and Wei, S.H.Y.: Topical fluoride therapy. *In* Pediatric Dentistry. Edited by R.E. Stewart, K.C. Barber, T.K. Troutman, and S.H.Y. Wei. St. Louis, C.V. Mosby Co., 1982.
4. Wei, S.H.Y.: The potential benefits to be derived from topical fluorides in fluoridated communities. *In* International Workshop on Fluorides and Dental Caries Reduction. Edited by D. Forrester, and E. Schulz, Jr., University of Maryland, 1974, pp. 178–258.
5. Horowitz, H.S.: A review of systemic and topical fluorides for the prevention of dental caries. Community Dent. Oral Epidemiol., *1*:104–114, 1973.
6. Bibby, B.G., and Turesky, S.S.: A note on the duration of caries inhibition produced by fluoride applications. J. Dent. Res., *26*:105–108, 1947.
7. Syrrist, A., and Karlsen, K.: A five-year report on the effect of topical application of sodium fluoride on dental caries experience. Br. Dent. J., *97*:1–6, 1954.
8. Sundvall-Hagland, I.: Sodium fluoride application to the deciduous dentition: A clinical study. Acta Odontol. Scand., *13*:5–14, 1955.
9. Houwink, B., Backer-Dirks, O., and Kwant, G.W.: A nine-year study of topical application with stannous fluoride in identical twins and caries experience five years after ending the application. Caries Res., *8*:27–38, 1974.
10. Horowitz, H.S., and Kau, M.C.: Retained anticaries protection from topically applied acidulated phosphate fluoride: 30- and 36-month post-treatment effects. J. Prev. Dent., *1*:22–27, 1974.
11. Averill, H.M., Averill, J.E., and Ritz, A.G.: A two-year comparison of three topical fluoride agents. J. Am. Dent. Assoc., *74*:996–1001, 1967.
12. Horowitz, H.S., and Heifetz, S.B.: Evaluation of topical applications of stannous fluoride to teeth of children born and reared in a fluoridated community: Final report. J. Dent. Child., *26*:355–361, 1969.
13. DePaola, P.F., and Mellberg, J.R.: Caries experience and fluoride uptake in children receiving semi-annual prophylaxis with an acidulated phosphate fluoride paste. J. Am. Dent. Assoc., *87*:155–159, 1973.
14. Downer, M.C., Holloway, P.J., and Davies, T.G.H.: Clinical testing of a topical fluoride preventive programme. Br. Dent. J., *141*:242–247, 1976.
15. Shannon, I.L., and Edmonds, E.J.: Enamel solubility reduction by acidulated phosphate fluoride (APF) treatment. Community Dent. Oral Epidemiol., *6*:12–16, 1978.
16. Shannon, I.L., and Edmonds, E.J.: Effect of pH and fluoride concentration on enamel solubility reduction by APF solutions. Dent. Hyg., *52*:231–235, 1978.
17. Dijkman, A.G., Tak, J., and Arends, J.: Comparison of fluoride uptake by human enamel from acidulated phosphate fluoride gels with different fluoride concentrations. Caries Res., *16*:197–200, 1982.
18. Sluiter, J.S., and Purdell-Lewis, D.J.: Lower fluoride concentration for topical application. Caries Res., *18*:56–62, 1984.

19. Hagan, P., Rozier, G., and Bawden, J.W.: Caries preventive effects of full- and half-strength topical acidulated phosphate fluoride. J. Dent. Res., *63*:772, 1984.
20. Wefel, J.S., and Wei, S.H.Y.: In vitro evaluation of fluoride uptake from a thixotropic gel. Pediatr. Dent., *1*:97–100, 1979.
21. Wei, S.H.Y., and Connor, C.J., Jr.: Fluoride uptake and retention in vivo following topical fluoride applications. J. Dent. Res., *62*:830–832, 1983.
22. Cobb, B.H., Rozier, G.R., and Bawden, J.W.: A clinical study of the caries preventive effects of an APF solution and APF thixotropic gel. Pediatr. Dent., *2*:263–266, 1980.
23. Shern, R.J., Duany, L.F., Senning, R.S., and Zinner, D.D.: Clinical study of an amine fluoride gel and acidulated phosphate fluoride gel. Community Dent. Oral Epidemiol., *4*:133–136, 1976.
24. DePaola, P.F., Soparkar, M., Van Leeuwen, M., and DeVelis, R.: The anticaries effect of single and combined topical fluoride systems in school children. Arch. Oral Biol., *25*:649–653, 1980.
25. Horowitz, H.S., et al.: Evaluation of self-administered prophylaxis and supervised toothbrushing with acidulated phosphate fluoride. Caries Res., *8*:39, 1974.
26. Heifetz, S.B., Horowitz, H.S., and Driscoll, W.S.: Two-year evaluation of self-administered procedure for the topical application of acidulated phosphate fluoride: Final report. J. Public Health Dent., *30*:7, 1970.
27. Mellberg, J.R., et al.: Short intensive topical APF applications and dental caries in a fluoridated area. Community Dent. Oral Epidemiol., *6*:117–120, 1978.
28. Zahran, M.: Effect of topically applied acidulated phosphate fluoride on dental caries. Community Dent. Oral Epidemiol., *4*:240–243, 1976.
29. Mainwaring, P.J., and Naylor, M.N.: A three-year clinical study to determine the separate and combined caries inhibiting effects of sodium monofluorophosphate toothpaste and an acidified phosphate fluoride gel. Caries Res., *12*:202–212, 1978.
30. Haupt, M., Koenigsberg, S., and Shey, Z.: The effect of prior toothcleaning on the efficacy of topical fluoride treatment. Clin. Prev. Dent., *5*:8–10, 1983.
31. Carter, W.J., et al.: The effect of topical fluoride on dental caries experiences in adult females of a military population. J. Dent. Res., *34*:73–76, 1955.
32. Harris, N.O., et al.: Stannous fluoride topically applied in aqueous solution in caries prevention in a military population. USAF Technical Documentary Report No. SAM-RDR-64-26, 1964.
33. Klinkenberg, E., and Bibby, B.G.: The effect of topical applications of fluorides on dental caries in young adults. J. Dent. Res., *29*:4–7, 1950.
34. Kutler, B., and Ireland, R.L.: The effect of sodium fluoride on the dental caries experience in adults. J. Dent. Res., *32*:458–462, 1953.
35. Muhler, J.C.: The effect of a single topical application of stannous fluoride on the incidence of dental caries in adults. J. Dent. Res., *37*:415, 1958.
36. Muhler, J.C., et al.: The arrestment of incipient dental caries in adults after the use of the three different forms of SnF_2 therapy; results after 30 months. J. Am. Dent. Assoc., *75*:1401–1407, 1967.
37. Rickles, H.N., and Becks, H.: The effects of an acid and a neutral solution of sodium fluoride on the incidence of dental caries in young adults. J. Dent. Res., *30*:757–765, 1951.
38. Scola, F.D., and Ostrom, C.A.: Clinical evaluation of stannous fluoride when used as a constituent of a compatible prophylactic paste, as a topical solution, and in a dentifrice in naval personnel. J. Am. Dent. Assoc., *77*:594–597, 1968.
39. Arnold, F.A., Dean, H.T., and Singleton, D.E.: The effect on caries incidence of a single topical application of a fluoride solution to the teeth of young males of a military population. J. Dent. Res., *23*:155–162, 1944.
40. Swango, P.A.: The use of topical fluorides to prevent dental caries in adults: A review of the literature. J. Am. Dent. Assoc., *107*:447–450, 1983.
41. Silverstone, L.M., et al.: Remineralization of natural and artificial lesions in human dental enamel in vitro. Caries Res., *15*:138–157, 1981.
42. Goodman, S.D.: An in vitro model to assess the interproximal gel coverage and fluoride uptake following a topical fluoride application. M.S. Thesis, University of Iowa, 1983.

Chapter 3
STANNOUS FLUORIDE IN CLINICAL DENTISTRY

Norman Tinanoff

There is growing evidence that stannous fluoride (SnF_2) has antiplaque properties, even at mouthrinse concentrations. Initially these antiplaque effects were considered non-specific; however, recent findings have shown that *Streptococcus mutans,* the oral bacteria most associated with dental caries, is reduced to a greater extent than other oral microorganisms. Stannous fluoride appears to affect the growth and adherence properties of bacteria rather than being bactericidal. Factors found to be important for stannous fluoride's antiplaque properties are: frequency of use, concentration, and stability of the commercial product. Research is necessary to further correlate the noted antiplaque properties of stannous fluoride with the clinical parameters of caries increment and gingivitis.

Fluoride (F), in various compounds and forms, has been the most successful agent in preventing dental caries. The mechanisms by which this ion reduces caries have traditionally been ascribed to its physico-chemical interaction with enamel. Recent research, however, has also been directed at the effect of fluoride compounds on bacterial metabolism. It is well established that the fluoride ion can alter bacterial metabolism at low concentrations and is bactericidal at higher concentrations. Yet, there now appears to be growing evidence that a specific fluoride compound, stannous fluoride, has greater antimicrobial effects than sodium fluoride (NaF). These antimicrobial effects of stannous fluoride appear to favorably affect clinical parameters such an plaque scores, gingival scores, and caries increment.

Most of the data regarding the antibacterial effects of stannous fluoride have been developed since 1975. Since there has not been sufficient time for many of these findings to be synthesized or replicated by different investigators, the research community has been cautious in accepting the potentially important antibacterial properties of stannous fluoride. Interestingly, many clinicians have become aware of the initial favorable research findings and have started using stannous fluoride as an antibacterial agent in their practices. The purpose of a symposium, such as the present one, is for researchers to consolidate and evaluate recent scientific findings objectively. This "state of the art" information should then be rapidly disseminated to practicing dentists, thus avoiding possible fads or swings of the pendulum in dental treatments.

The goal of this review will be to interface with the researcher and the clinician.

Therefore, this review will examine the current evidence of the antibacterial properties of stannous fluoride concentrating on its specific effect against *Streptococcus mutans*, the bacterium associated with dental caries. (Other contributors to this symposium discuss the effects of stannous fluoride on pathogenic organisms associated with gingivitis and certain forms of periodontal disease.) Furthermore, factors that determine the antimicrobial properties of stannous fluoride, such as frequency of use, compound stability, concentration, and commercial formulations, will also be addressed.

NONSPECIFIC ANTIPLAQUE EFFECTS OF STANNOUS FLUORIDE

Numerous animal and clinical studies demonstrate that stannous fluoride reduces dental plaque. König[1] in 1959 was the first to report plaque inhibition in rats when 0.1% SnF_2 was applied once a day for 35 days to their molars. Since there was not a sodium fluoride group in this study, it was not possible to tell whether the noted effects were due to the tin or fluoride ions.[1] Subsequent studies established that the antimicrobial effects were seen with stannous fluoride, but not with sodium fluoride. Andres et al. showed that stannous fluoride used daily as a mouthrinse decreased the number of salivary bacteria by 99%, whereas, sodium fluoride at the same fluoride ion concentration had no effect.[2] The difference in plaque accumulation was also reported in one subject rinsing with 100 ppm F as NaF or SnF_2. By electron microscopic criteria, plaque formation was drastically reduced when this subject rinsed with stannous fluoride. The few bacteria that were present on the subjects' enamel were not aggregated, suggesting that stannous fluoride interfered with bacterial adhesion or cohesion.[3]

Demonstration of this ability of stannous fluoride to reduce plaque formation has been impressively replicated. One single application of 8% SnF_2 was found to reduce plaque weight and visual plaque scores in 25 children.[4] Other clinical studies have found that toothpastes containing stannous fluoride have antiplaque properties.[5,6,7] In an experiment comparing twice daily rinsing with 0.2 or 0.3% SnF_2 mouthrinse, it was found that SnF_2 was comparable in antiplaque properties to 0.1% chlorhexidine $(C_{22}H_{30}Cl_2N_{10})$.[8] Also, twice daily rinsing with 0.04% SnF_2 or 0.1% SnF_2 has been shown to significantly reduce plaque scores in short-term studies.[9,10]

Using plaque area scores as a criterion for nonspecific plaque reduction is probably not ideal with a study involving stannous fluoride. Stannous fluoride enhances pellicle deposition on teeth,[11,12] which can be visually interpreted as plaque, especially if disclosing solutions are used to identify deposits on teeth (Fig. 3–1A, B). This nonbacterial deposit can greatly increase in thickness in areas that are not brushed well, or if the patient is not using regular toothpastes that contain abrasives (Fig. 3–2).

As the pellicle deposits increase in thickness they may appear yellow or light brown, due to the tin in the pellicle reacting with sulfides found in the mouth.[13] The extrinsic tooth staining as a result of the use of dilute stannous fluoride preparations is minimal, but it could potentially bias a clinical trial since examiners may not be able to remain "blind."

SPECIFIC EFFECTS AGAINST *STREPTOCOCCUS MUTANS*

The bacterium, *Streptococcus mutans*, is essential in the initiation of smooth surface caries and the augmentation of fissure caries.[14] A relatively high level of this organism in the mouth is correlated with increased caries increment,[15,16] and subjects

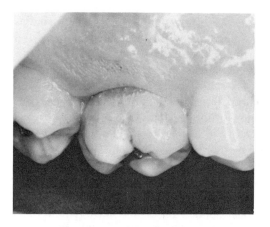 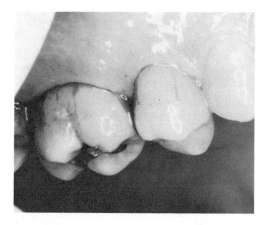

Fig. 3–1. Deposits on a first permanent molar visualized by disclosing solution in a subject who refrained from active oral hygiene for 1 week while rinsing twice daily with (a) placebo, or (b) 0.04% SnF_2.

with over 200,000 *S. mutans*/ml saliva have been shown to be at risk for development of new carious lesions.[17] Reduction of the oral flora with nonspecific antibacterial agents (e.g., antiseptics and antibiotics), has been shown to affect caries activity. Patients on prophylactic penicillin treatment for medical reasons have been noted to have reduced caries activity.[18] Chlorhexidine, used as a nonspecific plaque inhibitor, has also been shown to prevent the development of incipient lesions.[19] It seems more biologically acceptable, however, that specific suppression or elimination of *S. mutans* without reducing nonpathogenic bacteria would be superior to a nonspecific chemotherapeutic approach in preventing dental caries.

The first reported use of stannous fluoride to eliminate *Streptococcus mutans* was by Keene et al., in 1976.[20] These investigators found a reduction of *S. mutans* in five subjects after being treated with a prophy paste containing 9% SnF_2 and a 15-sec application of a 10% SnF_2 solution. They later showed that a 10% SnF_2 solution delivered with dental floss to interproximal sites significantly reduced *S. mutans*

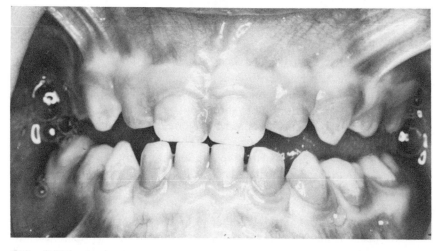

Fig. 3–2. Pellicle visualized with disclosing solution in a subject who brushed solely with a 0.4% SnF_2 gel.

more than a floss-saline treatment.[21] Unfortunately, it is not clear from these studies whether this was a specific reduction of *S. mutans,* since these investigations did not look at alterations in other aspects of the oral microflora.

A truly selective reduction of *Streptococcus mutans* by stannous fluoride was first documented in a study of 22 rampant caries subjects who rinsed twice a day with either acidulated sodium fluoride or stannous fluoride mouthrinses adjusted to 200 ppm F. Those subjects rinsing with stannous fluoride after 1,3, and 6 months had significant reductions of salivary *S. mutans,* while salivary total colony forming units and salivary lactobacilli were not affected by the stannous fluoride rinsing.[22] Svangberg and Rölla also found in 11 subjects who rinsed twice a day with 0.2% SnF_2 that *Streptococcus sanguis* and *S. mutans* were more reduced in plaque samples than in the total colony-forming units.[23] This selective effect of stannous fluoride against *S. mutans* was again noted with professional application of 8% SnF_2. This topical treatment reduced *S. mutans* in plaque and saliva, but no effect was seen on *S. sanguis* or lactobacilli.[24] Furthermore, 0.4% SnF_2 gel applied for 5 to 10 min/day in 12 cancer patients showed that this treatment prevented the expected rise in *S. mutans* associated with the radiation therapy. In three of these patients, *S. mutans* fell to nondetectable levels. Stannous fluoride, however, had no effect on lactobacillus levels.[25]

This specific suppression of *Streptococcus mutans* by stannous fluoride has been the recent focus of our clinical studies. We have continued to follow rampant caries patients who rinsed with either 200 ppm F as NaF or SnF_2. After 2 years, the group rinsing with stannous fluoride continued to have selective suppression of *S. mutans.* The stannous fluoride group has also shown significant lower caries incidence than those rinsing with sodium fluoride, which may be a reflection of the lower *S. mutans* levels.[26] In another study, we have examined the effect of twice daily brushing with 0.4% SnF_2 gel in an elderly institutionalized population. Brushing for 3 weeks with stannous fluoride reduced salivary *S. mutans* levels by 75 times without altering the salivary total colony-forming units.[27] Longer term studies are necessary in such a population to observe whether such a suppression of *S. mutans* would reduce the rate of root caries. Finally, we have explored the use of controlled release as a method of delivery of stannous fluoride. Subjects who had a molar tooth temporarily restored with stannous fluoride incorporated into polycarboxylate cement had a reduction in *S. mutans* over a 2-week period with no change in *Streptococcus sanguis* or total colony-forming units.[28] Table 3–1 summarizes those studies that have reported selective reductions of *S. mutans.*

In vitro studies have been done to explore how stannous fluoride affects *Streptococcus mutans*. Growth and adherence changes in *S. mutans* have been associated with the accumulation of tin in these cells.[29] Stannous fluoride concentrations above 125 ppm F have been shown to be bactericidal against *S. mutans,* and low concentrations of stannous fluoride (10 ppm F) have produced alterations in *S. mutans* DNA and glucan production.[30] However, the reason why the antimicrobial properties of stannous fluoride affect primarily *S. mutans* is still unclear. One explanation is that mouthrinsing with stannous fluoride inhibits acid formation in plaque for several hours, and that this subsequent increase in plaque pH could create an ecologic disadvantage for *S. mutans*.[23] Another proposed explanation is that tooth surfaces that have been disinfected with an antimicrobial agent such as stannous fluoride are more easily recolonized with *Streptococcus sanguis* because of its greater oral reservoir.[24]

Table 3–1. Studies Reporting Selective Reduction in *S. mutans* by the Use of SnF$_2$ Preparations

Investigators	# of Subjects using SnF$_2$	Length of Study	Frequency	Concentration of SnF$_2$ (%)	Mode of Delivery
Tinanoff et al., 1982	12 (rampant caries)	6 months	2x/day	.08	mouthrinse
Svangberg and Rölla, 1982	11	3 days	2x/day	.2	mouthrinse
Svangberg and Westergren, 1983	8	1 month	3x	8.0	professional application
Klock and Tinanoff, 1984	12 (rampant caries)	2 years	2x/day	.08	mouthrinse
Tinanoff et al., 1984	7	2 weeks	continuous	(0.3 ppm)	controlled release
Keene et al., 1984	12 (cancer patients)	1 year	1x, 5–10 min	.4	applicator trays
Potter et al., 1984	14 (elderly institutionalized)	3 weeks	2x/day	.1	brush-on

In vitro studies comparing the growth altering potential of stannous fluoride on *S. mutans,* lactobacilli, *S. sanguis, S. mitis,* etc. are indicated to examine how stannous fluoride preferentially affects different oral strains and what specific trait of *S. mutans* makes it susceptible to stannous fluoride.

CLINICAL FACTORS IMPORTANT FOR THE ANTIPLAQUE EFFECTS OF STANNOUS FLUORIDE

Frequency of Use. Frequency of use of an oral antimicrobial agent is related to its substantivity, i.e., the prolonged association of the material to oral surfaces. Many antimicrobial agents have been tried in attempts to reduce plaque. However, the oral retention of most of these compounds is only transient, and consequently they produce little plaque inhibition. Chlorhexidine is an exception, since it is strongly adsorbed to oral surfaces, creating a reservoir of antibacterial acitivity.[31] It appears that stannous fluoride is also retained in the mouth for prolonged periods. In vitro, *Streptococcus mutans* cells have been shown to take up large quantities of tin after exposure to stannous fluoride[29] with the antibacterial effects of stannous fluoride being directly proportional to tin retention.[32] Clinically, tin retention in plaque of subjects rinsing with 0.2% SnF$_2$ has been reported to be 40% after 7 hours.[33]

It seems clear that the substantivity of dilute stannous fluoride requires this agent to be used frequently for antibacterial properties. Those studies that have observed antiplaque effects of stannous fluoride have used either one application of 8 to 10% SnF$_2$, or twice daily exposure to more dilute (0.04 to 0.4%) SnF$_2$ solutions. No antibacterial effects of stannous fluoride were apparent in a study that used only once-a-day exposures to 0.1%.[34] To examine the relationship of frequency of use of stannous fluoride to its effectiveness on the oral microflora, we have compared brushing once vs twice daily with 0.4% SnF$_2$ on 17 subjects. When the subjects brushed twice daily they showed large reductions in salivary *Streptococcus mutans*, while once daily brushing was not nearly as effective (Fig. 3–3). These studies suggest that the substantivity of dilute stannous fluoride requires this agent to be used twice a day for antibacterial properties.

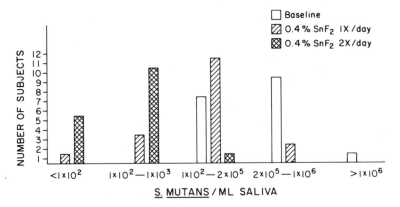

Fig. 3–3. Seventeen subjects ranked by *S. mutans*/ml saliva at baseline, when brushing once a day with 0.4% SnF$_2$, or when brushing twice a day with 0.4% SnF$_2$.

Stability. Stannous fluoride has long been known to undergo hydrolysis in an aqueous environment, resulting in the conversion of stannous ions to insoluble stannous hydroxide. The clouding sometimes noted in an aqueous solution of stannous fluoride is probably the stannous hydroxide precipitate. The instability of stannous fluoride can be reduced by not altering the native pH of the solution (pH 3.2 for 0.4% SnF$_2$), not adding foreign ions to the solution (e.g., abrasives as in toothpastes, buffers, etc.), or by not making a solution very dilute. The stability of stannous fluoride can be greatly improved by storing the solution in plastic bottles, and by adding glycerine or other water-insoluble materials to diminish the activity of free stannous ion, thereby reducing the rate of hydrolysis.[35]

We have examined the effect of the pH of a stannous fluoride solution on its activity as an antibacterial agent. Table 3–2 demonstrates the results of an experi-

Table 3–2. Effect of Altering the pH of SnF$_2$ or NaF (250 ppm F) on Antibacterial Parameters against *S. mutans*

	Agent pH	Terminal pH of broth[†]	Plaque score[*]	Dry Plaque weight (mg)	Sn/mg plaque (μg)
NaF	2.0	5.70	4	6.4 ± 0.9	ND[ss]
	6.0	5.09	4	6.5 ± 0.3	ND
SnF$_2$	2.0	7.32	<1	1.8 ± 0.1	15.8 ± 2.9
	3.0	7.26	<1	2.4 ± 0.5	36.8 ± 7.9
	4.0	7.03	1	2.6 ± 0.5	33.3 ± 4.4
	5.0	6.51	3	5.7 ± 0.4	20.1 ± 0.4
	6.0	5.87	4	5.9 ± 0.8	3.6 ± 0.7

[†]Initial pH 7.5
[*]Scored by McCabe method
[ss]None detected
N = 3; x̄ ± S.D.

ment in which *Streptococcus mutans* growth on stainless steel wires was used as a model to evaluate bacterial inhibition (Fig. 3–4 demonstrates this model system). As shown by visual plaque scores and dry plaque weight, the bacteria on the wires exposed to 0.1% SnF_2 at or below pH 4 were most affected by intermittent fluoride exposures. Stannous fluoride at pH 5 and 6 and sodium fluoride at both pH levels had little effect on bacterial growth. Also apparent from this experiment was the association of increased tin deposits in the bacteria (Sn/mg plaque) with greater antibacterial properties of stannous fluoride. This experiment clearly demonstrates that maintaining the appropriate pH of stannous fluoride in solution is critical for its stability and its consequent antibacterial properties.

Concentration. The effectiveness of stannous fluoride in reducing bacterial colonization of enamel also appears to be proportional to its concentration. In vitro studies have shown statistically greater antibacterial properties of stannous fluoride at 0.1% than at 0.04%.[29] The role of concentration of stannous fluoride has also been observed clinically. Toothpastes containing 0.14% SnF_2 have shown plaque-inhibiting properties, whereas toothpaste containing 0.03% have shown no effect.[5] The most convincing demonstration of the effect of concentration of stannous fluoride, however, was that shown by Svatun et al.[8] In this study, 12 subjects rinsed twice a day with either a placebo, 0.1% chlorhexidine ($C_{22}H_{30}Cl_2N_{10}$), 0.2% SnF_2, or 0.3% SnF_2. Mean plaque index scores were: 0.35 with the 0.2% SnF_2; 0.20 with the 0.3% SnF_2; 0.12 with $C_{22}H_{30}Cl_2N_{10}$; 1.02 with placebo. To optimize the antiplaque effects of stannous fluoride, most recent studies have increased the concentration of SnF_2 to 0.4%, the same fluoride ion level as found in fluoride toothpastes.

Commercial Formulation. To produce a stable stannous fluoride, manufacturers make a nonaqueous solution by incorporating stannous fluoride into anhydrous glycerine at temperatures around 300° F. The solution may be thickened with binders, and various coloring and flavoring agents are also added to suit consumer preferences. Apparently the processing is difficult to control, since there is large variability among manufacturers, methods and even among batches from the same source (Table 3–3). Possible problems in manufacturing include high temperatures, which may hydrolyze the tin or volatilize the fluoride, not absolutely water-free processing, poor quality control, and ignorance of necessary conditions (e.g., pH and temperature). These errors in manufacturing may produce products that do not have full antibacterial properties (Fig. 3–4). One company has patented its manufacturing process, which it claims produces a more stable product (Brand F in Table 3–3).

Table 3–3. The pH, and the Percent of Theoretical Fluoride or Stannous Ions of Six Commercial 0.4% SnF_2 Gels

	BRAND A	BRAND B	BRAND C	BRAND D	BRAND E	BRAND F
Expiration Dates	8/85–11/85	7/86–8/87	11/84–2/85	12/85	11/86–1/87	4/86–6/86
pH 1:1	3.8–4.1	4.1–4.5	3.2–3.3	4.2–4.5	4.5–4.8	2.9–3.1
F^- %	92–96	85–93	97–99	99–106	95–100	102–117
Sn^{++} %	21–51	49–77	69–82	85–91	85–95	99–102

Fig. 3–4. *S. mutans* plaque formation on stainless steel wires exposed to "Brand E" (left) or "Brand F" (center) SnF$_2$ gels diluted to 0.1%, or water control (right). "Brand E" with elevated pH and less stannous ion allows some bacterial accumulations on the wire.

CONCLUDING REMARKS

Use of stannous fluoride preparations in dentistry appears to be strongly revitalized due to its apparent antibacterial properties. This compound has been part of dental practice for many years without any reported adverse effects. The tin ion, although a heavy metal, is known to be safe even when ingested. An average diet contains

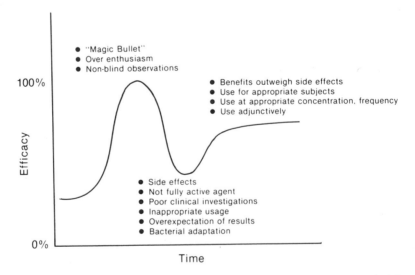

Fig. 3–5. Characteristic pattern of acceptance of many new drugs. Early enthusiasm is followed by disappointments until appropriate uses are established.

hundreds of milligrams of tin/day with the greatest source being from canned foods. The fluoride concentration in 0.4% SnF_2 is equal to that of fluoridated toothpastes, a level which is generally regarded as safe.

Side effects such as staining and metallic aftertaste could potentially deter some patient acceptance. The impression we have received in our clinical trials, however, suggests that many patients subjectively report a "clean mouth feeling" when using stannous fluoride, which apparently negates any complaints.

Certainly, there is more research to be conducted with regard to the antimicrobial effect of stannous fluoride. In vitro and in vivo studies are needed to further explain how and to what extent, stannous fluoride produces a selective suppression of the apparent pathogenic oral bacteria. Clinical studies must also continue to document the long-term effects of this agent on clinical parameters, and the oral conditions that would benefit most from stannous fluoride therapy. Further understanding of the appropriate concentrations, frequencies, delivery systems, and adjunctive use of stannous fluoride with other treatment regimens will enable the most effective and efficient use of this agent in the prevention of caries and periodontal disease (Fig. 3–5).

REFERENCES

1. König, K.G.: Dental caries and plaque accumulation in rats treated with stannous fluoride and penicillin. Helv. Odontol. Acta, 6:40–44, 1959.
2. Andres, C.J., Shaffer, J.C., and Windeler, A.S., Jr.: Comparison of the antibacterial properties of stannous fluoride and sodium fluoride mouthwashes. J. Dent. Res., 53:457–460, 1974.
3. Tinanoff, N., Brady, J.M., and Gross, A.: The effect of NaF and SnF_2 mouth-rinses on bacterial colonization of tooth enamel: TEM and SEM studies. Caries Res., 10:415–426, 1976.
4. Caldwell, P.E., Crawford, J.J., Hicks, E.P., and Stanmeyer, W.R.: Topical stannous fluoride effect on the adherence of dental plaque. AADR Abstract #576, 1977.
5. Svatun, B.: Plaque-inhibiting effect of dentifrices containing stannous fluoride. Acta Odontol. Scand., 36:205–210, 1978.
6. Bay, I., and Rölla, G.: Plaque inhibition and improved gingival condition by use of a stannous fluoride toothpaste. Scand. J. Dent. Res., 88:313–315, 1980.
7. Ogaard, B., Gjermo, P., and Rölla, G.: Plaque-inhibiting effect in orthodontic patients of a dentifrice containing stannous fluoride. Am. J. Orthod., 78:266–271, 1980.
8. Svatun, B., Gjermo, P., Eriksen, H.M., and Rölla, G.: A comparison of the plaque-inhibiting effect of stannous fluoride and chlorhexidine. Acta Odontol. Scand., 35:247–250, 1977.
9. Tinanoff, N., Hock, J., Camosci, D., and Hellden, L.: Effect of stannous fluoride mouthrinse on dental plaque formation. J. Clin. Periodontol., 7:232–241, 1980.
10. White, S.T., and Taylor, P.P.: The effect of stannous fluoride on plaque scores. J. Dent. Res., 58:1850–1852, 1979.
11. Tinanoff, N., and Weeks, D.B.: Current status of SnF_2 as an antiplaque agent. Pediatr. Dent., 1:199–204, 1979.
12. Ellingsen, J.E., Eggen, K.H., and Rölla, G.: Surface properties of hydroxyapatite treated with NaF or SnF_2. IADR Abstract #435, Dallas, 1974.
13. Vogel, R.I.: Intrinsic and extrinsic discoloration of the dentition—a literature review. J. Oral Med., 30:99–104, 1975.
14. Tanzer, J.M.: Essential dependence of smooth surface caries on, and augmentation of fissure caries by, sucrose and *Streptococcus mutans* infection. Infect. Immun., 25:526–531, 1979.
15. Klock, B., and Krasse, B.: Effect of caries-preventive measures in children with high numbers of *S. mutans* and lactobacilli. Scand. J. Dent. Res., 86:221–230, 1978.
16. Köhler, B., Pettersson, B., and Bratthall, D.: *Streptococcus mutans* in plaque and saliva and the development of caries. Scand. J. Dent. Res., 89:19–25, 1981.
17. Maltz, M., Zickert, I., and Krasse, B.: Effect of intensive treatment with chlorhexidine on the number of *Streptococcus mutans* in saliva. Scand. J. Dent. Res., 89:445–449, 1981.
18. Handelman, S.L., Mills, J.R., and Hawes, R.R.: Caries incidence in subjects receiving long term antibiotic therapy. J. Oral Ther. Pharm., 2:338–345, 1966.
19. Löe, H., von der Fehr, F.R., and Rindom-Schiott, C.: Inhibition of experimental caries by plaque prevention. The effect of chlorhexidine mouthrinses. Scand. J. Dent. Res., 80:1–9, 1972.

20. Keene, H.J., Shklair, I.L., and Hoerman, K.C.: Partial elimination of *Streptococcus mutans* from selected tooth surfaces after restoration of carious lesions and SnF_2 prophylaxis. J. Am. Dent. Assoc., *93*:382–383, 1976.
21. Keene, H.J., Shklair, I.L., and Mickel, G.J.: Effect of multiple dental floss-SnF_2 treatment on *Streptococcus mutans* in experimental plaque. J. Dent. Res., *56*:21–27, 1977.
22. Tinanoff, N., Manwell, M.A., Camosci, D.A., and Klock, B.: Microbiologic effect of SnF_2 vs NaF mouthrinse after 6 months. IADR Abstract #517, New Orleans, 1982.
23. Svanberg, M., and Rölla, G.: *Streptococcus mutans* in plaque and saliva after mouthrinsing with SnF_2. Scand. J. Dent. Res., *90*:292–298, 1982.
24. Svanberg, M., and Westergren, G.: Effect of SnF_2 administered as a mouthrinse or topically applied on *Streptococcus mutans, Streptococcus sanguis,* and lactobacilli in dental plaque and saliva. Scand. J. Dent. Res., *91*:123–129, 1983.
25. Keene, H.J., Fleming, T.J., Brown, L.R., and Dreizen, S.: Lactobacilli and *S. mutans* in cancer patients using fluoride gels. IADR Abstract #429, Dallas, 1984.
26. Klock, B., and Tinanoff, N.: Effect of SnF_2 on different microorganisms and caries incidence in adults. IADR Abstract #331, Dallas, 1984.
27. Potter, D.E., et al.: SnF_2 as an adjunct to toothbrushing in an elderly institutionalized population. Special Care in Dentistry, (in press).
28. Tinanoff, N., Seigrist, B., and Lang, N.P.: Safety and antibacterial properties of controlled release SnF_2. IADR Abstract #332, Dallas, 1984.
29. Tinanoff, N., and Camosci, D.A.: Microbiological, ultrastructural, and spectroscopic analyses of the anti-tooth-plaque properties of fluoride compounds in vitro. Arch. Oral Biol., *25*:531–543, 1980.
30. Ferretti, G.A., Tanzer, J.M., and Tinanoff, N.: The effect of fluoride and stannous ions on *Streptococcus mutans*: Viability, growth, acid, glucan production, and adherence. Caries Res., *16*:298–307, 1982.
31. Rölla, G., Löe, H., and Schiott, C.R.: The affinity of chlorhexidine for hydroxyapatite and salivary mucins. J. Periodont. Res., *5*:90–95, 1970.
32. Camosci, D.A., and Tinanoff, N.: Anti-bacterial determinants of stannous fluoride. J. Dent. Res., (in press).
33. Attramadal, A., and Svatun, B.: Uptake and retention of tin by *S. mutans*. Acta Odontol. Scand., *38*:349–354, 1980.
34. McHugh, W.D., Eisenberg, D.H., Leverett, D.H., and Jensen, O.E.: Microbial plaque composition after daily rinsing with SnF_2 and NaF. IADR Abstract #204, Sydney, 1983.
35. Hefferren, J.J.: Qualitative and quantitative tests for stannous fluoride. J. Pharm. Sci., *52*:1090–1096, 1963.

Chapter 4

THE ROLES OF PROPHYLAXES AND DENTAL PROPHYLAXIS PASTES IN CARIES PREVENTION

Louis W. Ripa

In this chapter, the roles of prophylaxes and dental prophylaxis pastes in caries prevention are reviewed. There are several procedures in which self-administered or professionally administered prophylaxes have been advocated for dental caries prevention; although there is little clinical evidence to substantiate these methods, and it has been concluded that dental prophylaxis pastes play no direct role in caries prevention. While the presence of fluoride in some prophylaxis pastes implies a cariostatic benefit, the effectiveness of these pastes has not been established. The fluoride in prophylaxis pastes, however, may be useful in replenishing the enamel fluoride lost from the surfaces of teeth during polishing.

Dental prophylaxis pastes contain a variety of abrasive materials that are used to clean the clinical crowns of the teeth. These prophylaxis pastes also remove extrinsic stains, salivary pellicle, and bacterial plaque from tooth surfaces. Because they are abrasive, these pastes can remove small amounts of tooth structure, thereby polishing the enamel surface.[1]

Dental prophylaxis pastes are available in fluoride or fluoride-free varieties. Because of the etiologic relationship between bacterial plaque and both gingivitis and caries, dental prophylaxis pastes have been studied in programs designed to control gingival disease, dental caries, or both. They have also been employed in professional- and self-treatment programs.

This review is concerned only with the role of dental prophylaxis in caries inhibition. Therefore, the following subjects will be addressed:

(1) The caries inhibitory potential of *frequent* professional prophylaxis with fluoride-free or fluoride-containing prophylaxis pastes.

(2) The caries inhibitory potential of *infrequent* professional prophylaxis with fluoride-free or fluoride-containing prophylaxis pastes.

(3) The caries inhibitory potential of self-applied fluoride-containing prophylaxis pastes.

(4) The routine use of fluoride-containing or fluoride-free prophylaxis pastes prior to professional topical fluoride treatment.

FREQUENT PROFESSIONAL PROPHYLAXIS

Dental caries develops beneath a layer of acidogenic bacterial plaque. Elimination of this dental plaque from the surfaces of the teeth should result in caries inhibition. This rationale was the basis for a series of studies begun in Sweden in the 1970s, which involved frequent professional tooth cleaning (Table 4–1).

The first of these studies, by Lindhe and Axelsson, involved school children in Karlstad, Sweden.[2–5] Subsequently, four other programs[6–10] were initiated by Lindhe and Axelsson, one of which involved adults.[9,10] The "Karlstad model" was the forerunner of at least 12 other studies designed to evaluate the effectiveness of frequent professional prophylaxis.[11–23] Most of these investigations were conducted in Sweden,[11–14,20,23] Norway,[17] and Denmark;[15,16,19] however, clinical trials were also con-

Table 4–1. Results of Clinical Trials of Frequent Professional Prophylaxis

Study	Initial Age of Subjects (Years)	Duration (Years)	Frequency of Professional Prophylaxis	Caries Surface Increment		Difference	
				Control	Experimental	No. Surfaces	%
Lindhe & Axelsson[2] (1973)	7–14	1	1×/2 weeks	3.25	0.06	3.19	98
Axelsson & Lindhe[3] (1974)	7–14	2	1×/2 weeks	5.60	0.19	5.41	97
Lindhe et al.[4] (1975)	7–14	3	1×/4 weeks 1×/8 weeks	8.44	0.45	7.99	95
Axelsson & Lindhe[5] (1977)	7–14	4	1×/8 weeks	10.11	0.74	9.37	93
Axelsson et al.[6] (1976)	13–14	2	1×/2 weeks	6.4 4.8	1.6 2.2	4.8 2.6	75 54
Axelsson & Lindhe[7] (1975)	13–14	1	1×/2 weeks[a]	0.26	0.70	−0.44	—
Axelsson & Lindhe[8] (1981)	13–14	1.5	1×/2 weeks[b]	N.A.	N.A.	N.A.	N.A.
Axelsson & Lindhe[9] (1978)	Adults	3	1×/8 weeks[c] 1×/12 weeks	2.51	0.04	2.47	98
Axelsson & Lindhe[10] (1981)	Adults	6	1×/12 weeks[c]	13.94	0.22	13.72	98
Hamp et al.[11] (1978)	10–11	3	1×/3 weeks	12.8	6.3	6.5	51
Hamp & Johansson[12] (1982)	16–19	3	1×/3 weeks[d]	3.3	1.0 1.2 2.0	2.3 2.1 1.3	70 64 39
Badersten et al.[13] (1975)	10–12	1	1×/4 weeks	4.5	3.4	1.1	24
Klock & Krasse[14] (1978)	9–12	2	1×/2 weeks[e] 1×/4 weeks 1×/4 weeks	4.28 4.28 2.76	0.92 1.46 1.64	3.36 2.82 1.12	79 66 41

Table 4–1. *(continued)* Results of Clinical Trials of Frequent Professional Prophylaxis

Study	Initial Age of Subjects (Years)	Duration (Years)	Frequency of Professional Prophylaxis	Caries Surface Increment		Difference	
				Control	Experimental	No. Surfaces	%
Poulsen et al.[15] (1976)	7	1	1×/2 weeks	1.42	0.43	0.99	70
Agerbaek et al.[16] (1978)	7	2	1×/3 weeks[f]	2.09	1.40	0.69	33
Kjaerheim et al.[17] (1980)	7–13	2	1×/2 weeks	1.61	0.57	1.04	65
Bellini et al.[18] (1981)	7	4	1×/4 weeks	1.78	0.81	0.97	54
Vestergaard et al.[19] (1978)	5–13	2	1×/2 weeks	1.75	1.44	0.31	18
Gisselsson et al.[20] (1983)	10–11	2	1×/3 weeks	4.5	1.7	2.8	62[g]
Ashley & Sainsbury[21] (1981)	11	3	1×/2 weeks	4.66	4.97	−0.31	—
Craig et al.[22] (1981)	11–12	2	1×/2 weeks	2.9	2.6	0.3	10
Zickert et al.[23] (1982)	13–14	2	1×/4 weeks	5.4	3.2	2.2	41
			1×/4 weeks	5.4	3.2	2.2	41
			1×/12 weeks	7.0	3.8	3.2	46
			1×/12 weeks	7.0	4.2	2.8	40

[a]Both groups received professional prophylaxis; the control group was treated with a fluoride-free paste and the experimental group with a fluoride-containing paste

[b]Half-mouth treatments provided to control and experimental groups

[c]Treatment frequency 1×/8 weeks for 1st two years; thereafter 1×/12 weeks

[d]Three treatment groups: Group A received professional prophylaxis 1×/3 weeks for 1st year, 1×/4 weeks for 2nd year, and 2×/52 weeks for 3rd year; Group B received professional prophylaxis 1/×3 weeks for 1st year, 2×/52 weeks for 2nd and 3rd years; Group C received professional prophylaxis 1×/3 weeks for 1st year with no treatments in 2nd and 3rd years

[e]Three treatment and two control groups based upon *S. mutans* and lactobacillus counts

[f]Treatment frequency 1×/2 weeks for 1st year and 1×/3 weeks for 2nd year

[g]This reduction did not maintain itself in a 2-year post-treatment evaluation reported in the same article

ducted in England,[21,24] New Zealand,[22] and Brazil.[18] (For additional non-English language citations, see the review by Bellini and coworkers.[18])

Lindhe and Axelsson's original study[2–5] employed specially trained dental nurses who treated the children in their schools during the academic year. The program involved a combination of preventive procedures. Meticulous dental cleaning was performed by the dental nurses using rubber and bristle cups rotated by a dental engine. During the cleaning, a topical application of fluoride was administered by the incorporation of 5% Na_2PO_3F in the abrasive paste. Oral hygiene instruction was provided which included supervision in the Bass method of toothbrushing and interdental flossing. Participating children received treatments every 2 weeks during the first 2 years of the study. During the last 2 years, the frequency of treatments was reduced to 4-week and 8-week intervals. At the end of 4 years, Axelsson and Lindhe reported a caries increment of 0.74 DS/child in the treatment group and 10.11 DS/child in the control group. The difference of 9.37 in the 4-year increment between the two groups represented a caries reduction of 93%.[5] In a subsequent study by Axelsson and coworkers,[6] a similar treatment plan was employed in some groups of children, while in others the professional prophylaxis was omitted and replaced by topical application of 0.5% chlorhexidine gel. The chlorhexidine-treated children had higher caries increments than observed for the professional prophylaxis groups, suggesting the importance of repeated professional tooth cleaning in the caries inhibition obtained. In a study on Swedish adults, Axelsson and Lindhe reported 6-year caries increments of 0.22 surfaces in the treated group and 13.72 surfaces in the control group, a difference of 98%.[9,10] In this study, the control group received no special oral hygiene care, whereas the treated group received professional tooth cleaning with a fluoride-containing paste at intervals of 8 and 12 weeks.

Other investigators have either obtained statistically significant but smaller caries reductions than those obtained by Axelsson and Lindhe, or have failed to achieve a statistically significant result.[11–23]

Viewed collectively, and without regard to program particulars, the studies listed in Table 4–1 suggest that frequent professional prophylaxis can significantly reduce dental caries incidence in both children and adults. In many of these studies, however, a multiplicity of preventive modalities including supervised toothbrushing and interproximal flossing, oral hygiene instruction, and dietary advice were provided to the participants, which made it impossible to isolate the effects specific to the professional toothcleaning. Moreover, in the Scandinavian studies, fluoride contact generally was either included in the treatment regimen or was simultaneously provided in school-based public health programs in which the children were also participating. Thus, the results of mechanical plaque removal cannot be isolated from the chemical effects that fluoride might have had on the teeth.

In the original Karlstad study, a prophylaxis paste containing 5% Na_2PO_3F was used.[2–5] Additionally children in the control group participated in a supervised school-based program involving brushing with a 0.2% solution of sodium fluoride. Since this program had been in effect for 10 years, both the control and treated children, at least in the upper grades, were exposed to the fluoride brushing program before the professional prophylaxis program began. In the study by Hamp and coworkers,[11] two different fluoride-containing prophylaxis pastes were used (one in the first year, the other during the next 2 years), and the children in both the experimental and control groups rinsed with a 0.2% NaF mouthrinse. Furthermore, all of the children participated in the fluoride mouthrinsing program before the professional prophylaxis program began. In the studies reported by Poulsen,[15] Agerbaek,[16] Kjaerheim,[17] Badersten,[13] Vestergaard,[19] and their coworkers, the children participated in school-based fluoride rinse or brushing programs during the course of the study. The older children had been participants of the fluoride programs before being treated in the prophylaxis program.[17] In the study by Kjaerheim and coworkers in Norway[17] and Klock and Krasse in Sweden,[14] the children had sealants placed on the occlusal surfaces of their teeth either before or during the studies, which raises the additional question of the possible influence of sealants in the other studies from these countries.

When evaluating the effectiveness of frequent professional prophylaxis on caries inhibition, it is important to realize that in those studies that did not employ a fluoride-containing prophylaxis paste and/or in which the children were not participating in other fluoride methods, statistically significant differences in the caries increments between control and treatment groups were not achieved.[15,16,21,22,24]

Even if the results of these studies could be construed as unequivocal, there are two considerations which discourage both the continued study of this method and its implementation. These considerations are the declining prevalence of dental caries and the cost of delivering a service involving as many as 20 professional cleanings a year.

The average initial caries prevalence scores of the children in the Axelsson-Lindhe studies listed in Table 4–1 were high, indicating populations that were very susceptible to decay. In Norway, Sweden, and Denmark, where frequent professional prophylaxis has been studied, the caries prevalence has decreased by 27%[25] to approximately 50%[26] or more.[27] In addition, the number of caries-free children has doubled in at least two of these countries.[25,27] The results of Badersten and coworkers[13] and of Klock and Krasse[14] indicated that the subjects with the highest caries risk benefited the most from the program. Conversely, the studies by Vestergaard and coworkers,[19] Ashley and Sainsbury,[21,24] and Craig and coworkers[22] employed populations with relatively low caries prevalences, and the high frequency professional treatments failed to achieve a significant effect. It is conceivable, therefore, that with the decline in caries prevalence, the results of new studies would prove less dramatic. Since the procedure is a labor-intensive one, the cost of the program would be expected to be high relative to other school-based programs in which many children can be treated simultaneously or to office-based programs in which individual children are treated less frequently.[28] Thus, the potential for a lower "yield" when treating low-risk children and higher costs when targeted high-risk children are treated make this an impractical method of caries prevention.

INFREQUENT PROFESSIONAL PROPHYLAXIS

Annual or biannual prophylaxes routinely have been provided to patients as a dental preventive measure. The usefulness of infrequent professional toothcleanings on dental caries, however, is doubtful. In 1966, Bibby stated, "Evidence that dental prophylaxis . . . prevents dental decay is lacking."[29] Subsequently, Ripa and coworkers demonstrated that 4 biannual professionally administered prophylaxes had no effect on caries.[30]

In the 1940s, attempts were made to use a prophylaxis paste as a vehicle for topical fluoride. The idea of incorporating fluoride into a prophylaxis paste was attractive because the paste proved to be an effective cariostatic agent, a prophylaxis of the teeth could then be substituted for the more time-consuming two-step prophylaxis-topical fluoride treatment which was then being advocated and subsequently adopted.

Bibby and coworkers[31] were the first to test the preceding concept. They used a pumice-hydrogen peroxide slurry containing 1% NaF. After 1 year, caries reductions of 25 and 42% from two and three applications respectively were reported.[31] A second study by Bibby, however, failed to confirm these findings.[32]

No clinical studies of fluoride-containing prophylaxis pastes were conducted in the 1950s, possibly because of the incompatibility of many abrasives with fluoride.[33] In the 1960s, the clinical effectiveness of professionally administered stannous fluoride-containing prophylaxis pastes with either a silex or lava pumice abrasive system were clinically tested[34–40] (Table 4–2). Later, studies of other systems,[41,42] including APF-containing prophylaxis pastes were conducted.[43–46]

Although in the United States prophylaxis pastes have been marketed that contain APF, SnF_2, NaF, and Na_2PO_3F, clinical studies of their effectiveness have been

Table 4–2. Results of Clinical Trials of Professionally Administered Fluoride-Containing Prophylaxis Pastes

Study	Initial Age of Subjects (Years)	Agent & Abrasive	Duration (Years)	No. Applications/ Year	% Caries Inhibition DMFT	DMFS
FLUORIDE-DEFICIENT COMMUNITIES						
Bibby et al.[31] (1946)	6–14	1% NaF, pumice	1	2	—	25
	6–14	1% NaF, pumice	1	3	—	42
Bibby[32] (1948)	6–15	1% NaF, pumice	1	3	—	0
Peterson et al.[34] (1963)	10–13	17.5% SnF$_2$, silex	2	2	35(39)	42(34)[a]
Scola & Ostrom[35] (1966)	17–24	17.5% SnF$_2$, lava pumice	1	1	12	12
Scola & Ostrom[36] (1968)	17–24	17.5% SnF$_2$, lava pumice	2	1	26	12
Bixler & Muhler[37] (1964)	5–18	8.9% SnF$_2$, lava pumice	1	2	31	35
Bixler & Muhler[38] (1966)	5–18	8.9% Snf$_2$, lava pumice	2	2	30	34
			3	2	33	35
Horowitz & Lucye[40] (1966)	8–10	8.9% SnF$_2$, lava pumice	2	1	+7.9	+5.9
Peterson et al.[41] (1969)	10–13	2% KF, H$_3$PO$_4$, lava pumice	2	1	14(12)	16(15)
Szwejda[42] (1971)	6–10	SnF$_2$[b]	2	1	13	17
			3	1	17	19
			2	1	9	20
			3	1	19	20

meager. Three studies have been reported on the APF paste when professionally applied. One study showed a marginal caries reduction[44] and the others reported no significant clinical effect to the permanent[45] or primary[46] teeth. With the exception of a brief abstract,[47] there has been only one reported clinical trial in which the marketed stannous fluoride paste was tested in a professionally administered program.[48] In that study, the paste was not found to be effective when applied at 6-month intervals for 3 years. This paste, which used a zirconium silicate abrasive system, has since been withdrawn from the market.

The display of clinical results in Figure 4–1 suggests a relationship between the

Table 4–2. *(continued)*   Results of Clinical Trials of Professionally Administered Fluoride-Containing Prophylaxis Pastes

Study	Initial Age of Subjects (Years)	Agent & Abrasive	Duration (Years)	No. Applications/ Year	% Caries Inhibition	
					DMFT	DMFS
FLUORIDE-DEFICIENT COMMUNITIES						
Szwejda[43] (1972)	6–10	1.23% APF, sodium meta-phosphate[c]	1	1	30	20
			2	—	20	18
			3	—	23	18
DePaola & Mellberg[44] (1973)	10–13	1.23% APF, SiO_2	2	2	17	21
Barenie et al.[45] (1976)	9–14	1.23% APF, SiO_2	2	2	+7(+8)	+5(+8)
Beiswanger et al.[48] (1980)	8–16	9% SnF_2, $ZrSiO_4$	1	2	8	17[d]
			2	2	22	20
			3	2	17	15
OPTIMALLY FLUORIDATED COMMUNITIES						
Gish & Muhler[39] (1965)	6–14	8.9% SnF_2, lava pumice	1	2	29(45)	40(42)
Peterson et al.[41] (1969)	11–13	2% KF, H_3PO_4, lava pumice	2	1	19(8)	15(12)
Schutze et al.[46] (1974)	3–5	1.23% APF, SiO_2	1	3		+16[e]

[a] Two independent examiners
[b] No other information given about the composition of this paste
[c] Prophylaxis followed by APF rinse
[d] Water F fluctuated from 0.0 to 0.8 ppm
[e] Defs primary teeth
NaF = sodium fluoride; SnF_2 = stannous fluoride; KF = potassium fluoride; APF = acidulated phosphate fluoride; SiO_2 = silicon dioxide; $ZrSiO_4$ = zirconium silicate

frequency of treatment and caries inhibition; nevertheless, it is apparent that this method of topical fluoride application gives inconsistent results. Infrequent applications of highly concentrated fluoride agents, such as fluoride-containing prophylaxis pastes, are believed to exert their cariostatic influence by reacting with the enamel to produce a more acid-insoluble surface layer. Enamel fluoride uptake from in vivo application has been demonstrated with the use of fluoride-containing prophylaxis pastes, especially APF pastes.[44,45,49,51] The amount of fluoride acquired by enamel from prophylaxis pastes, however, has been less than that found when aqueous solutions or gels were used.[49,50] In addition, the fluoride acquired from a prophylaxis paste was lost from the teeth before the next 6-month treatment.[51] As discussed by Mellberg,[52] the ingredients of a prophylaxis paste formulation influence the potential effectiveness of fluoride more than any other vehicle associated with topical fluoride

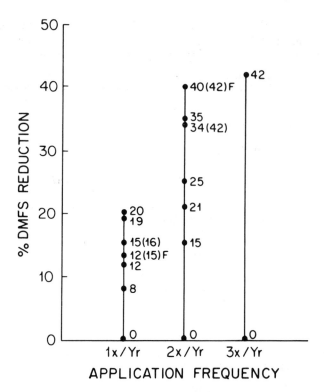

( ) = Results of 2nd independent examiner
F = Fluoridated community
Only final results of studies are listed

Fig. 4–1. Although there is a trend toward greater caries reductions with more applications, it is evident from the array of the data that infrequent professional treatments with fluoride-containing prophylaxis pastes give inconsistent results.

treatments. It is not surprising, therefore, that the results of clinical studies of fluoride-containing prophylaxis pastes are unpredictable. Based upon the results of the studies listed in Table 4–2, a fluoride-containing prophylaxis paste cannot be recommended as the sole agent when performing a topical fluoride application.

SELF-APPLIED PROPHYLAXIS

Self-applied methods of topical fluoride application are desirable for caries prevention because costs are lower compared to professionally applied methods and more children can be treated more often than when an operator is involved.

Generally, two types of fluoride containing prophylaxis pastes have been clinically tested in programs in which the pastes have been brushed on the teeth by the subjects themselves. These are a 9% SnF_2 paste with zirconium silicate abrasive (SnF_2-$ZrSiO_4$) and a 1.23% APF paste utilizing a silicon dioxide abrasive (APF-SiO_2).

Table 4–3 lists the results of these studies on primary[53] and permanent teeth,[54–64] in optimally fluoridated[57,65,66] and fluoride-deficient communities.[53–64] (Results of other studies using the SnF_2-$ZrSiO_4$ paste have been reported in incomplete form and are not listed in Table 4–3.[67,68]) The results of these studies range from no effect on caries to a maximum caries inhibition of 64%. The clinical results of the studies

Table 4–3. Results of Clinical Trials of Self-Administered Fluoride-Containing Prophylaxis Pastes

Study	Initial Age	Agent & Abrasive	Duration (Years)	No. Applications/Year	% Caries Inhibition DMFT	DMFS
FLUORIDE-DEFICIENT COMMUNITIES						
Muhler et al.[54] (1970)	6–14	9% SnF$_2$, ZrSiO$_4$	1	1	41	64
Fleming et al.[55] (1976)	11–15	9% SnF$_2$, ZrSiO$_4$	2	1	+3	+16
			2	2	24	5
Muhler[59] (1976)	7–13	SnF$_2$-alkali, aluminum silicate	1	1	31	33
Horowitz & Bixler[56] (1976)	9–14	9% SnF$_2$, ZrSiO$_4$	3	2 (1st year) 1 (2nd & 3rd years)	14(20)	15(23)[a]
Gish et al.[57] (1975)	6–14	9% SnF$_2$, ZrSiO$_4$	3	2	32	37
Woodhouse[58] (1978)	12–13	10% SnF$_2$, ZrSiO$_4$	3	2	19	16
Mellberg et al.[60] (1974)	6–14	1.23% APF, SiO$_2$	3	2	14	21
Gray et al.[64] (1980)	10	1.23% APF, ? abrasive[b]	2	2	24	23
Ringleberg et al.[63] (1976)	5–9	2.2% APF, pumice	2	2		25
Woods et al.[53] (1976)	5–9	10% SnF$_2$, ZrSiO$_4$	2	3	36	50
			2	3	76 (dft)	
Long[61] (1972)	10–13	1.23% APF, SiO$_2$	2	3	17	
Trubman & Crellin[62] (1973)	8.2	1.23% APF, SiO$_2$	3	4	4	10
		1.23% APF, SiO$_2$[c]	3	4	25	22
OPTIMALLY FLUORIDATED COMMUNITIES						
Gunz[65] (1971)	7–11	9% SnF$_2$, ZrSiO$_4$	1.2	1	11	11
Lang et al.[66] (1970)	6–14	9% SnF$_2$, ZrSiO$_4$	1.5	2	27(41,41)	38(42,42)[d]
Gish et al.[57] (1975)	6–14	9% SnF$_2$, ZrSiO$_4$	3	2	30	25

[a]Two independent examiners
[b]Followed by 1% NaF rinse
[c]Followed by APF gel-tray application
[d]Three independent examiners
SnF$_2$ = stannous fluoride; APF = acidulated phosphate fluoride; ZrSiO$_4$ = zirconium silicate; SiO$_2$ = silicon dioxide

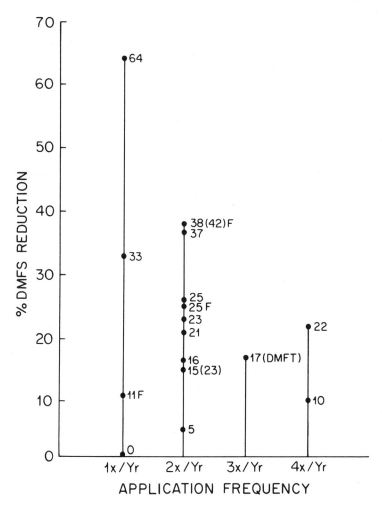

( ) = Results of 2nd independent examiner
F = Fluoridated community
Only final results of studies are listed

Fig. 4–2. There is little uniformity in the clinical results achieved by the self-administration of fluoride-containing prophylaxis pastes.

listed in Table 4–3 are also presented in Figure 4–2. Approximately two-thirds of the caries reductions from the use of self-applied fluoride-containing prophylaxis pastes are 25% or less. While a principal difference between the studies is the number of times (1 to 4 times/year) in which the subjects applied the paste, the brushing frequency does not seem to have influenced the results. In fact, scrutiny of Table 4–3 and Figure 4–2 reveals little uniformity or predictability in the caries inhibition achieved. Consequently, self-brushing with a fluoride-containing prophylaxis paste cannot be regarded as a reliable method of caries prevention.

In the 1960s and 1970s, a number of "brush-in" programs were conducted in schools and other locations where people congregated.[67-77] In addition, the United States Navy developed a three-part caries preventive program for their personnel that included self-brushing with a fluoride-containing prophylaxis paste.[78] The "brush-in" program as a community measure for caries control is waning.[79] This is not only because the results of self-brushing studies have been equivocal, but because supervision of large groups as they self-brush is cumbersome compared to the logistically simpler mouthrinsing method. In addition, there were complaints of nausea when the SnF_2-$ZrSiO_4$ treatment paste was used.[69] The withdrawal of the SnF_2-$ZrSiO_4$ paste from the market probably also contributed to the demise of the "brush-in" method for community caries prevention.

PRIOR TO PROFESSIONAL TOPICAL FLUORIDE TREATMENT

Fluoride-Containing Pastes

There have been eight clinical studies in which a fluoride containing prophylaxis paste has been used for preparatory tooth cleaning when a topical fluoride solution was professionally administered[34–40,48,80,81] (Table 4–4). Two of these studies were conducted in fluoridated communities;[39,81] the others, in fluoride-deficient ones.[34–38,40,48,80,81]

The results of the studies listed in Table 4–4 suggest that the two fluoride-containing topical agents used on the same patient can provide greater protection than the use of a fluoride-containing prophylaxis paste alone. However, six of the eight studies did not include a treatment group in which a topical application of aqueous fluoride was preceded by a prophylaxis with a fluoride-free paste.[35–39,81–83] Thus, these studies could not determine whether a fluoride-containing prophylaxis paste used in conjunction with an aqueous topical fluoride application could provide better caries protection than professional application of topical fluoride preceded by a fluoride-free paste. Of the two studies that were designed to answer this question, neither found the combination of a professional prophylaxis performed with a fluoride-containing paste and a topical fluoride application to be superior to a topical fluoride treatment that was preceded by a prophylaxis with a fluoride-free paste.[40,48] In one of these studies, however, none of the three treatment groups derived any benefits from the various fluoride methods.[40]

Thus, clinical studies do not support the conclusion that a fluoride-containing prophylaxis paste used immediately before a professional topical fluoride treatment will provide additive cariostatic benefits.

Fluoride-Free Pastes

Based upon the initial clinical studies with sodium fluoride,[82,83] it has been recommended that topical applications of neutral sodium fluoride, stannous fluoride, and acidulated phosphate fluoride be preceded by a professional cleaning of the teeth.[84–86] While several clinical trials of professional topical fluoride applications were conducted in which a professional prophylaxis of the teeth was not performed,[87–90] there were no clinical studies, until recently, that tested the need for the prior prophylaxis.

Within the last decade, in vivo[91–94] and in vitro[95–97] studies have evaluated the permeability of plaque and pellicle to fluoride and have examined fluoride absorption by uncleaned enamel. These studies have demonstrated that organic integuments covering the tooth surface were permeable to the fluoride ion and did not significantly interfere with fluoride uptake by the enamel (see review by Ripa[98]).

Table 4–4. Results of Clinical Studies Using a Combination of Professionally Administered Fluoride-Containing Prophylaxis Pastes and Aqueous Topical Fluoride Solutions

Study	Initial Age	Fluoride Treatments	Duration (Years)	No. Applications/ Year	% Reduction DMFT	% Reduction DMFS
FLUORIDE-DEFICIENT COMMUNITIES						
Bixler and Muhler[37] (1964)	5–18	P	1	2	31(39)	35(34)[a]
		P + T		2	46(40)	48(42)
Bixler and Muhler[38] (1966)		P	2	2	24(36)	31(37)
		P + T			35(46)	39(43)
		P	3	2	29(36)	29(38)
		P + T			39(58)	40(56)
Scola and Ostrom[35] (1966)	17–24	P	1	1	12	12
		P + T			47	43
Scola and Ostrom[36] (1968)		P	2	1	26	12
		P + T			34	27
Peterson et al.[34] (1963)	10–13	P	2	1	28	34
		P + T			29	32
		P	2	2	35	42
Horowitz & Lucye[40] (1966)	8–11	P	2	1	+8	+6
		T			+2	+8
		P + T			+3	1
Downer et al.[80] (1976)	11–12	P + T + D[b]	3	2	34	31
Beiswanger et al.[48] (1980)	8–14	P	3	2	17	15
		T			47	52
		P + T			45	52
OPTIMALLY FLUORIDATED COMMUNITIES						
Gish and Muhler[39] (1965)	6–14	P	1	2	29(45)	40(42)
		P + T			68(58)	75(58)
Muhler et al.[81] (1967)	dental students	P + T + D	2.5	2	68	64

[a]Two independent examiners
[b]Na$_2$PO$_3$F is the active ingredient in D; APF the active ingredient in P and T; in all other studies, SnF$_2$ is the active ingredient in P, T, and D
P = fluoride containing prophylaxis paste; T = professional topical fluoride application, D = fluoride containing dentifrice

Table 4-5. Mean DMFS Increments of Children Receiving Biannual Professional APF Gel-Tray Topical Fluoride Treatments

Treatment	Ripa et al.[102] (1983) 2 Years	Houpt et al.[104] (1983)* 2 Years	Katz et al.[105] (1984) 2.5 Years	Ripa et al.[103] (1984) 3 Years
Prophylaxis + Topical F	2.12	2.05	2.23	3.33
Self-Brush + Topical F	1.87	2.48	2.33	3.18
Topical F	2.02	2.14	2.09	3.19

*Two-year increment of a fourth nonparticipating group was 2.50

Spurred by these studies and by others that explored the fluoride-plaque-enamel surface interrelationship,[99-101] three independent clinical trials were begun to assess the need for a prior prophylaxis when administering a topical fluoride treatment.[102-105] All three studies involved professionally administered biannual topical fluoride treatments administered as gels in mouthtrays to school children. Each of the studies had three separate treatment groups that differed in the method of prior tooth cleaning. One group received a meticulous hygienist-administered prophylaxis with a fluoride-free or fluoride-containing prophylaxis paste followed by a 4-minute gel tray topical fluoride application. This group served as the positive control. A second group brushed and flossed their teeth under supervision using a fluoride-free dentifrice and then received the same gel-tray treatment. The third group received a gel-tray fluoride treatment without any prior cleaning of the teeth.

The 2- and 3-year results of these studies are presented in Table 4-5. Neither the type nor degree of cleaning had an effect on the clinical efficacy of the professional topical fluoride application. Therefore, based upon these studies, the prophylaxis step appears to be unnecessary as a *routine* procedure when performing a professional topical fluoride treatment.

CONCLUDING REMARKS

This review considered the various methods in which prophylaxis pastes have been employed in programs to limit dental caries. Since these pastes are intended to clean and polish the teeth, any caries inhibition that accrues from their use should be associated with the mechanical removal of plaque from the enamel surface. Because prophylaxis pastes may contain fluoride, a second potential cariostatic mechanism is the chemical alteration of enamel solubility.

Listed in Table 4-6 are the different procedures that utilize prophylaxis pastes for caries prevention. The most recent is a program of frequent professional tooth cleaning, as often as every 2 weeks. This method has been studied extensively since it was first reported in 1973. Nevertheless, because of the confounding variables in the study designs of many of the trials and the lack of positive findings in others, the results must be considered equivocal. Because frequent professional prophylaxis appears to be an approach that is most useful in high caries risk subjects and it is labor intensive, the contemporary combination of decreasing caries rates and increasing fiscal consciousness has rendered it prematurely obsolete.

Annual or biannual dental prophylaxes are routinely performed in most dental of-

Table 4–6. Evaluation of Procedures That Use Prophylaxis Pastes for Caries Prevention

Procedure	Comments
Frequent professional prophylaxis (up to $1 \times /2$ weeks) with a fluoride-containing or fluoride-free paste	Results equivocal Method impractical
Infrequent professional prophylaxis ($1 \times$ or $2 \times /$ year) with a fluoride-free paste	No evidence of effectiveness
Infrequent professional prophylaxis ($1 \times$ or $2 \times /$ year) with a fluoride-containing paste	No marketed product proven effective in repetitive controlled clinical trials
Self-administered prophylaxis with a fluoride-containing paste	No marketed product proven effective in repetitive controlled clinical trials
Preliminary prophylaxis with a fluoride-containing paste prior to professional topical fluoride application	Does not provide additive benefits to the topical fluoride treatment
Preliminary prophylaxis with a fluoride-free paste prior to professional topical fluoride application	Does not provide additive benefits to the topical fluoride treatment Unnecessary as a routine procedure

fices. At one time, they were the mainstay of many school-based preventive programs that employed hygienists. Surprisingly, there is no evidence to support that this infrequent procedure by itself provides any cariostatic benefit. Although fluoride has been included in some prophylaxis paste formulations in order to impart chemical resistance to the teeth during the cleaning procedure, the clinical evidence is insufficient to conclude that their infrequent application (once or twice a year) has a significant impact on caries. With one exception, none of the fluoride-containing prophylaxis pastes on the United States market have been tested for clinical efficacy in a program of professional application. In the single exception in which an APF paste was tested, one study was marginally positive[44] and the others reported no benefit from the use of this paste.[45,46] Similarly, clinical studies of the self-administration of this paste have produced conflicting results.[60–62]

Traditionally, a dental prophylaxis has always preceded the annual or biannual professional application of topical fluoride. Both fluoride-containing and fluoride-free pastes have been used for this purpose. Neither type of paste, when evaluated in controlled clinical trials, has been found to impart additive benefits to the topical fluoride procedure. Thus, the continued routine use of the prior prophylaxis step when performing a professional topical fluoride treatment cannot be recommended.

The conclusion of this review is that prophylaxis pastes play no direct role in the prevention of dental caries. This does not imply that prophylaxis pastes provide no dental benefit, since they do perform an esthetic function by removing extrinsic stains from the teeth and are an important component of the methodology for the prevention of gingivitis and periodontal disease.[106]

The findings of this review regarding the cariostatic benefits of fluoride-containing prophylaxis pastes in both self-administered and professionally administered programs are consistent with the United States Food and Drug Administration's failure to recognize any fluoride-containing prophylaxis paste as therapeutic and the absence of these pastes from the ADA's list of accepted therapeutic agents.[107]

Because of the scarcity of evidence supporting the cariostatic benefits of fluoride-containing prophylaxis pastes, their presence on the market may be questioned. The answer to this question must consider that mechanical cleaning of the teeth using prophylaxis pastes is also performed for reasons unrelated to caries prevention. Whenever teeth are polished, a thin layer of enamel is abraded, resulting in the loss of fluoride from the tooth surface.[108-111] While the clinical significance of this loss has never been studied, conventional wisdom deems the loss to be undesirable. Fluoride-containing prophylaxis pastes may replenish the fluoride abraded by the polishing procedure.[51] The presence of fluoride in these pastes does not harm and may do some good, a sentiment that has been expressed by others.[48,112] Because the ingredients of a prophylaxis paste may restrict the availability of fluoride in its formulation,[52] each paste containing fluoride should be required to be tested for bioavailability. If bioavailability can be established, then the presence of fluoride in the paste is justified.

REFERENCES

1. Mellberg, J.R.: The relative abrasivity of dental prophylaxis pastes and abrasives on enamel and dentin. Clin. Prevent. Dent., 6:13–18, 1979.
2. Lindhe, J., and Axelsson, P.: The effect of controlled oral hygiene and topical fluoride applications on caries and gingivitis in Swedish school children. Community Dent. Oral. Epidemiol., 1:9–16, 1973.
3. Axelsson, P., and Lindhe, J.: The effect of a preventive programme on dental plaque, gingivitis, and caries in school children. Results after one and two years. J. Clin. Periodontol., 1:126–138, 1974.
4. Lindhe, J., Axelsson, P., and Tollskog, G.: Effect of proper oral hygiene on gingivitis and dental caries in Swedish school children. Community Dent. Oral Epidemiol., 3:150–155, 1975.
5. Axelsson, P., and Lindhe, J.: The effect of a plaque control program on gingivitis and dental caries in school children. J. Dent. Res., 56:C142–C148, 1977.
6. Axelsson, P., Lindhe, J., and Waseby, J.: The effect of various plaque control measures on gingivitis and caries in school children. Community Dent. Oral Epidemiol., 4:232–239, 1976.
7. Axelsson, P., and Lindhe, J.: Effect of fluoride on gingivitis and dental caries in a preventive program based on plaque control. Community Dent. Oral Epidemiol., 3:156–160, 1975.
8. Axelsson, P., and Lindhe, J.: Effect of oral hygiene instruction and professional tooth cleaning on caries and gingivitis in school children. Community Dent. Oral Epidemiol., 9:251–255, 1981.
9. Axelsson, P., and Lindhe, J.: Effect of controlled oral hygiene procedures on caries and periodontal disease in adults. J. Clin. Periodontol., 5:133–151, 1978.
10. Axelsson, P., and Lindhe, J.: Effect of controlled oral hygiene procedures on caries and periodontal disease in adults. Results after six years. J. Clin. Periodontol., 8:239–248, 1981.
11. Hamp, S.-E., et al.: Effect of a field program based on systematic plaque control of caries and gingivitis in school children after three years. Community Dent. Oral Epidemiol., 6:17–23, 1978.
12. Hamp, S.-E., and Johansson, L.-A.: Dental prophylaxis for youths in their late teens. I. Clinical effect of different preventive regimes on oral hygiene, gingivitis, and dental caries. J. Clin. Periodontol., 9:22–34, 1982.
13. Badersten, A., Egelberg, J., and Koch, G.: Effect of monthly prophylaxis on caries and gingivitis in school children. Community Dent. Oral Epidemiol., 3:1–4, 1975.
14. Klock, B., and Krasse, B.: Effect of caries-preventive measures in children with high numbers of *S. mutans* and lactobacilli. Scand. J. Dent. Res., 86:221–230, 1978.
15. Poulsen, S., et al.: The effect of professional toothcleansing on gingivitis and dental caries in children after one year. Community Dent. Oral Epidemiol., 4:195–199, 1976.
16. Agerbaek, N., Poulsen, S., Melsen, B., and Glavind, L.: Effect of professional toothcleansing every third week on gingivitis and dental caries in children. Community Dent. Oral Epidemiol., 6:40–41, 1978.
17. Kjaerheim, V., von der Fehr, F.R., and Poulsen, S.: Two-year study on the effect of professional tooth cleaning on children in Oppegard, Norway. Community Dent. Oral Epidemiol., 8:401–406, 1980.

18. Bellini, H.T., Arneberg, P., and von der Fehr, F.R.: Oral hygiene and caries: A review. Acta Odontol. Scand., *39*:257–265, 1981.
19. Vestergaard, V., Moss, A., Pedersen, H.O., and Poulsen, S.: The effect of supervised tooth cleansing every second week on dental caries in Danish school children. Acta Odontol. Scand., *36*:249–252, 1978.
20. Gisselsson, H., Bjorn, A.-L., and Birkhed, D.: Immediate and prolonged effect of individual preventive measures in caries and gingivitis susceptible children. Swed. Dent. J., *7*:13–21, 1983.
21. Ashley, F.P., and Sainsbury, R.H.: The effect of a school-based plaque control programme on caries and gingivitis. Br. Dent. J., *150*:41–45, 1981.
22. Craig, E.W., Suckling, G.W., and Pearce, E.I.F.: The effect of a preventive programme on dental plaque and caries in school children. N. Z. Dent. J., *77*:89–93, 1981.
23. Zickert, I., Lindvall, A.-M., and Axelsson, P.: Effect on caries and gingivitis of a preventive program based on oral hygiene measures and fluoride application. Community Dent. Oral Epidemiol., *10*:289–295, 1982.
24. Ashley, F.P., and Sainsbury, R.H.: Post-study effects of a school-based plaque control programme. Br. Dent. J., *153*:337–338, 1982.
25. Fejerskov, O., Antoft, P., and Gadegaard, E.: Decrease in caries experience in Danish children and young adults in the 1970s. J. Dent. Res., *61*:1305–1310, 1982.
26. von der Fehr, F.R.: Evidence of decreasing caries prevalence in Norway. J. Dent. Res., *6*:1331–1335, 1982.
27. Koch, G.: Evidence for declining caries prevalence in Sweden. J. Dent. Res., *61*:1340–1345, 1982.
28. Bagramian, R.A.: Oral hygiene procedures and pit and fissure sealants. *In* The Relative Efficiency of Methods of Caries Prevention in Dental Public Health. Edited by B.A. Burt. Ann Arbor, The University of Michigan Press, 1978, pp. 123–151.
29. Bibby, B.G.: Do we tell the truth about preventing caries? J. Dent. Child., *33*:269–279, 1966.
30. Ripa, L.W., Barenie, J.T., and Leske, G.S.: The effect of professionally administered biannual prophylaxes on the oral hygiene, gingival health, and caries scores of school children. J. Prev. Dent., *3*:22–26, 1976.
31. Bibby, B.G., Zander, H.A., McKelleget, M., and Labunsky, B.: Preliminary reports on the effect on dental caries of the use of sodium fluoride in a prophylactic cleaning mixture and in a mouthwash. J. Dent. Res., *25*:207–211, 1946.
32. Bibby, B.G.: Fluoride mouthwashes, fluoride dentifrices and other uses of fluorides in control of caries. J. Dent. Res., *27*:367–373, 1948.
33. Mellberg, J.R., and Nicholson, C.R.:In vitro evaluation of an acidulated phosphate fluoride prophylaxis paste. Arch. Oral. Biol., *13*:1223–1234, 1968.
34. Peterson, J.K., Jordan, W.A., and Snyder, J.R.: Effectiveness of stannous fluoride-silex-silicone prophylaxis paste. Northwest. Dent., *42*:276–278, 1963.
35. Scola, F.P., and Ostrom, C.A.: Clinical evaluation of stannous fluoride when used as a constituent of a compatible prophylactic paste, as a topical solution, and in a dentifrice in naval personnel. I. Report of findings after first year. J. Am. Dent. Assoc., *73*:1306–1311, 1966.
36. Scola, F.P., and Ostrom, C.A.: Clinical evaluation of stannous fluoride when used as a constituent of a compatible prophylactic paste, as a topical solution, and in a dentifrice in naval personnel. II. Report of findings after two years. J. Am. Dent. Assoc., *77*:594–597, 1968.
37. Bixler, D., and Muhler, J.C.: Effect on dental caries in children in a nonfluoride area of combined use of three agents containing stannous fluoride: A prophylactic paste, a solution, and a dentifrice. J. Am. Dent. Assoc., *68*:792–800, 1964.
38. Bixler, D., and Muhler, J.C.: Effect on dental caries in children in a nonfluoride area of combined use of three agents containing stannous fluoride: A prophylactic paste, a solution, and a dentifrice. II. Results at the end of 24 and 36 months. J. Am. Dent. Assoc., *72*:392–396, 1966.
39. Gish, C.W., and Muhler, J.C.: Effect on dental caries in children in a natural fluoride area of combined use of three agents containing stannous fluoride: A prophylactic paste, a solution, and a dentifrice. J. Am. Dent. Assoc., *70*:914–920, 1965.
40. Horowitz, H.S., and Lucye, H.: A clinical study of stannous fluoride in a prophylaxis paste and a solution. J. Oral Ther., *3*:17–25, 1966.
41. Peterson, J.K., Horowitz, H.S., Jordan, W.A., and Pugnier, V.: Effectiveness of an acidulated phosphate fluoride-pumice prophylactic paste: A two-year report. J. Dent. Res., *48*:346–350, 1969.
42. Szwejda, L.F.: Fluorides in community programs: Results after four years of study of various agents topically applied by two technics. J. Public Health Dent., *31*:166–176, 1971.
43. Szwejda, L.F.: Fluorides in community programs: A study for four years of the cariostatic effects of prophylactic pastes, rinses, and applications of various fluorides. J. Public Health Dent., *32*:110–118, 1972.
44. DePaola, P.F., and Mellberg, J.R.: Caries experience and fluoride uptake in children receiving

semiannual prophylaxes with an acidulated phosphate fluoride paste. J. Am. Dent. Assoc., *87*:155–159, 1973.

45. Barenie, J.T., et al.: Effect of professionally applied biannual applications of phosphate-fluoride prophylaxis paste on dental caries and fluoride uptake: Results after two years. J. Dent. Child., *43*:340–344, 1976.

46. Schutze, H.J., Jr., Forrester, D.J., and Balis, S.B.: Evaluation of a fluoride prophylaxis paste in a fluoridated community. Can. Dent. Assoc. J., *40*:675–683, 1974.

47. Peterson, J.K., Horowitz, H.S., Jordan, W.A., and Pugneir, V.: Effectiveness of acidulated phosphate fluoride and stannous zirconium hexafluoride in prophylactic pastes. IADR Programs and Abstracts of Papers, Abstract #277, March, 1967, p. 106.

48. Beiswanger, B.B., Mercer, V.H., Billings, R.J., and Stookey, G.K.: A clinical evaluation of a stannous fluoride prophylaxis paste and topical solution. J. Dent. Res., *59*:1386–1391, 1980.

49. Heifetz, S.B., Mellberg, J.R., Winter, S.J., and Doyle, J.: In vivo fluoride uptake by enamel of teeth of human adults from various topical fluoride procedures. Arch. Oral Biol., *15*:1171–1181, 1970.

50. Mellberg, J.R., Nicholson, C.R., and Trubman, A.: The acquisition of fluoride by tooth enamel in vivo from self-applied APF gel and prophylaxis paste. Caries Res., *7*:173–178, 1973.

51. Mellberg, J.R., Nicholson, C.R., Ripa, L.W., and Barenie, J.T.: Fluoride deposition in human enamel in vivo from professionally applied fluoride prophylaxis paste. J. Dent. Res., *55*:976–979, 1976.

52. Mellberg, J.R.: Chemistry of topical fluoride treatment. *In* Fluoride in Preventive Dentistry: Theory and Clinical Applications. Edited by J.R. Mellberg, and L.W. Ripa. Chicago, Quintessence Publishing Company, Inc., 1983, pp. 151–179.

53. Woods, R., Martin, N.D., and Barnard, P.D.: A community dental health project. I. Self-applied SnF_2-$ZrSiO_4$ prophylactic paste and dental caries in primary school children. Aust. Dent. J., *21*:205–210, 1976.

54. Muhler, J.C., et al.: The clinical evaluation of a patient-administered SnF_2-$ZrSiO_4$ prophylactic paste in children. I. Results after one year in the Virgin Islands. J. Am. Dent. Assoc., *81*:142–145, 1970.

55. Fleming, W.J., Burgess, R.C., and Lewis, D.W.: Effect on caries of self-application of a zirconium silicate paste containing 9 percent stannous fluoride. Community Dent. Oral Epidemiol., *4*:142–148, 1976.

56. Horowitz, H.S., and Bixler, D.: The effect of self-applied SnF_2-$ZrSiO_4$ prophylactic paste on dental caries: Santa Clara County, California. J. Am. Dent. Assoc., *92*:369–373, 1976.

57. Gish, C.W., Mercer, V.H., Stookey, G.K., and Dahl, L.O.: Self-application of fluoride as a community preventive measure: Rationale, procedures, and three-year results. J. Am. Dent. Assoc., *90*:388–397, 1975.

58. Woodhouse, A.D.: A longitudinal study of the effectiveness of self-applied 10 percent stannous fluoride paste for secondary school children. Aust. Dent. J., *23*:422–428, 1978.

59. Muhler, J.C.: A clinical evaluation of the dental caries experience in children receiving a self-applied stannous fluoride-alkali aluminum silicate prophylactic paste during a twelve-month study period. J. Dent. Child., *43*:345–346, 1976.

60. Mellberg, J.R., Peterson, J.K., and Nicholson, C.R.: Fluoride uptake and caries inhibition from self-application of an acidulated phosphate fluoride prophylaxis paste. Caries Res., *8*:52–60, 1974.

61. Long, J.G.: Self-applied fluoride paste: Effect on dental caries. J. Public Health Dent., *32*:161–164, 1972.

62. Trubman, A., and Crellin, J.A.: Effect on dental caries of self-application of acidulated phosphate fluoride paste and gel. J. Am. Dent. Assoc., *86*:153–157, 1973.

63. Ringleberg, M.L., Conti, A.J., and Webster, D.B.: An evaluation of single and combined self-applied fluoride programs in schools. J. Public Health Dent., *36*:229–236, 1976.

64. Gray, A.S., Gunther, D.M., and Munns, P.M.: Fluoride paste and rinse in a school dental program. Can. Dent. Assoc. J., *46*:651–654, 1980.

65. Gunz, G.M.: The effect of self-applied fluoride paste. J. Public Health Dent., *31*:177–181, 1971.

66. Lang, L.A., Thomas, H.G., Taylor, J.A., and Rothaar, R.E.: Clinical efficacy of a self-applied stannous fluoride prophylactic paste. J. Dent. Child., *37*:211–216, 1970.

67. Muhler, J.C.: Mass treatment of children with a stannous fluoride-zirconium silicate self-administered prophylactic paste for partial control of dental caries. J. Am. Coll. Dent., *35*:45–47, 1968.

68. Kelley, G.E.: Mass self-administered stannous fluoride prophylactic paste. Can. J. Public Health, *61*:226–231, 1970.

69. Muhler, J.C.: The clinical demonstration of the mass treatment of children with the SnF_2-$ZrSiO_4$ prophylactic paste. Initial observations concerning conduct of the study. J. Indiana Dent. Assoc., *47*:428–431, 1968.

70. Mercer, V.H., and Gish, C.W.: The self-administered stannous fluoride treatment paste for caries prevention in a community. J. Indiana Dent. Assoc., *47*:432–434, 1968.
71. Schimmele, R.G.: A suggested method for mass application of self-administered prophylactic paste using student auxiliary personnel. J. Indiana Dent. Assoc., *47*:435–436, 1968.
72. Kelly, G.E.: The Bloomington "brush-in." A new experience in dental caries prevention for mass treatment. J. Indiana Dent. Assoc., *48*:72–75, 1969.
73. Hoffman, E.: Operation "brush-in." J. S. Calif. Dent. Hyg. Assoc., *12*:14–15, 1970.
74. Foster, M.J.: Let's have a brush-in. J. Texas Dent. Hyg. Assoc., *7*:19, 1970.
75. Story, F.B.: North Carolina's first brush-in. J. NC Dent. Soc., *53*:15–17, 1970.
76. Smith, C.E.: "Brush-in": Self-applied topical fluorides. Ohio Dent. J., *44*:188–190, 1970.
77. Praven, J.R.: A "brush-in" with the self-administered stannous fluoride treatment paste. Texas Dent. J., *9*:10, 1973.
78. Unauthored: Preventive dentistry in the United States Navy. The three-agent program. J. Am. Dent. Assoc., *83*:994–995, 1971.
79. Horowitz, A.M., and Horowitz, H.S.: School-based fluoride programs: A critique. J. Prev. Dent., *6*:89–94, 1980.
80. Downer, M.C., Holloway, P.J., and Davies, T.G.H.: Clinical testing of a topical fluoride caries preventive programme. Br. Dent. J., *141*:242–250, 1976.
81. Muhler, J.C., Spear, L.B., Jr., Bixler, D., and Stookey, G.K.: The arrestment of incipient dental caries in adults after the use of three different forms of SnF_2 therapy: Results after 30 months. J. Am. Dent. Assoc., *75*:1402–1406, 1967.
82. Knutson, J.W., and Armstrong, W.D.: The effect of topically applied sodium fluoride on dental caries experience III. Report of findings for the third study year. Pub. Health Rep., *61*:1683–1689, 1946.
83. Knutson, J.W., Armstrong, W.D., and Feldman, V.D.: The effect of topically applied sodium fluoride on dental caries experience IV. Report of findings with two, four, and six applications. Pub. Health Rep., *62*:425–430, 1947.
84. Knutson, J.W.: Sodium fluoride solutions: Technic for application to the teeth. J. Am. Dent. Assoc., *36*:37–39, 1948.
85. Horowitz, H.S., and Heifetz, S.B.: The current status of topical fluorides in preventive dentistry. J. Am. Dent. Assoc., *81*:166–177, 1970.
86. Stookey, G.K., and Katz, S.: Chairside procedures for using fluorides for preventing dental caries. Dent. Clin. North Am., *16*:681–692, 1972.
87. Chrietzberg, J.E.: Toothbrushing as a substitute for quick cleansing in the topical fluoride treatment. J. Am. Dent. Assoc., *42*:435–438, 1951.
88. Cobb, H.B., Rozier, R.G., and Bawden, J.W.: A clinical study of the caries preventive effects of an APF solution and an APF thixotropic gel. Pediatr. Dent., *2*:263–266, 1980.
89. Shern, R. J., Duany, L.F., Senning, R.S., and Zinner, D.D.: Clinical study of an amine fluoride gel and acidulated phosphate fluoride gel. Community Dent. Oral Epidemiol., *4*:133–136, 1976.
90. Hass, R. L.: Effectiveness of a single application of stannous fluoride after toothbrushing. J. Am. Dent. Assoc., *71*:1391–1395, 1965.
91. Steele, R.C., Waltner, A.W., and Bawden, J.W.: The effect of tooth cleaning procedures on fluoride uptake in enamel. Pediatr. Dent., *4*:228–233, 1982.
92. Tinanoff, N., Wei, S.H.Y., and Parkins, F.M.: Effect of a pumice prophylaxis on fluoride uptake in tooth enamel. J. Am. Dent. Assoc., *88*:384–389, 1974.
93. Seppa, L.: Effect of dental plaque on fluoride uptake by enamel from a sodium fluoride varnish in vivo. Caries Res., *17*:71–75, 1983.
94. Bruun, C., and Stoltze, K.: In vivo uptake of fluoride by surface enamel of cleaned and plaque-covered teeth. Scand. J. Dent. Res., *84*:268–275, 1976.
95. Joyston-Bechal, S., Duckworth, R., and Braden, M.: The effect of artificially produced pellicle and plaque on the uptake of ^{18}F by human enamel in vitro. Arch. Oral Biol., *21*:73–78, 1976.
96. Klimek, J., Hellwig, E., and Ahrens, G.: Fluoride taken up by plaque, by the underlying enamel, and by clean enamel from three fluoride compounds in vitro. Caries Res., *16*:156–161, 1982.
97. Tinanoff, N., Wei, S.H.Y., and Parkins, F.M.: Effect of the acquired pellicle on fluoride uptake in tooth enamel in vitro. Caries Res., *9*:224–230, 1975.
98. Ripa, L.W.: Need for prior toothcleaning when performing a professional topical fluoride application. Review and recommendations for change. J. Am. Dent. Assoc., *109*:281–285, 1984.
99. Charlton, G., Blainey, B., and Schamschula, R.G.: Associations between dental plaque and fluoride in human surface enamel. Arch. Oral Biol., *19*:139–143, 1974.
100. McNee, S.G., Geddes, D.A.M., Main, C., and Gillespie, F.C.: Measurements of the diffusion coefficient of NaF in human dental plaque in vitro. Arch. Oral Biol., *25*:819–823, 1980.
101. Turtola, L.O.: Enamel microhardness and fluoride uptake underneath fermenting and nonfermenting artificial plaque. Scand. J. Dent. Res., *85*:373–379, 1977.

102. Ripa, L.W., Leske, G.S., Sposato, A., and Varma, A.: Effect of prior toothcleaning on biannual professional APF topical fluoride gel-tray treatments: Results after two years. Clin. Prevent. Dent., 5:3–7, 1983.
103. Ripa, L.W., Leske, G.S., Sposato, A., and Varma, A.: Effect of prior toothcleaning on biannual professional APF topical fluoride gel-tray treatments: Results after three years. Caries Res., (in press).
104. Houpt, M., Koenigsberg, S., and Shey, Z.: The effect of prior toothcleaning on the efficacy of topical fluoride treatment: Two-year results. Clin. Prevent. Dent., 5:8–10, 1983.
105. Katz, R.V., Meskin, L.H., Hensen, M.E., and Keller, D.: Topical fluoride and prophylaxis: A 30-month clinical trial. J. Dent. Res., 63:256 (Abstract 771), 1984.
106. Report of the Working Group on Preventive Dental Services: Preventive dental services: Practices, guidelines, and recommendations. Canada, Minister of Supply and Services, 1980, pp. 85–102.
107. Council on Dental Therapeutics and Council on Dental Materials, Instruments, and Equipment: Clinical products in dentistry. A desktop reference. J. Am. Dent. Assoc., 107:857–892, 1983.
108. Biller, I.R., Hunter, E.L., Featherstone, M.G., and Silverstone, L.M.: Enamel loss during a prophylaxis polish in vitro. J. Int. Assoc. Dent. Child., 11:7–12, 1980.
109. Stookey, G.K.: In vitro estimates of enamel and dentin abrasion associated with a prophylaxis. J. Dent. Res., 57:36, 1978.
110. Vrbic, V., Brudevold, F., and McCann, H.G.: Acquisition of fluoride by enamel from fluoride pumice pastes. Helv. Odontol. Acta, 11:21–26, 1967.
111. Zuniga, M.A., and Caldwell, R.C.: The effect of fluoride-containing prophylaxis pastes on normal and "white spot" enamel. J. Dent. Child., 36:345–349, 1969.
112. Report of the Working Group on Preventive Dental Services: Preventive dental services: Practices, guidelines, and recommendations. Canada. Minister of Supply and Services, 1980, p. 201.

Section II

CAPTAIN J. MICHAEL ALLEN, MODERATOR

Chapter 5
FLUORIDE SUPPLEMENTS AND DIETARY SOURCES OF FLUORIDE

Katherine Kula and Stephen H.Y. Wei

Fluoride supplements have been proven to cause definite caries reductions in children living in suboptimally fluoridated areas.

Although the mode of topical action for fluoride supplements is not well understood, fluoride tablets should remain in the mouth as long as possible prior to ingestion to produce high salivary fluoride levels. Since the caries preventive effects on teeth that erupt early during supplementation appear to decrease following discontinuation of supplements, topical fluoride should be prescribed following discontinuation.

The current fluoride supplement schedule approved by the American Dental Association (ADA) adjusts the fluoride dosage of children from birth to 2 years of age downward to 0.25 mg F/day in order to prevent the occurrence of fluorosis. Animal studies suggest that fluorosis is based on dose per kilogram of body weight, although prolonged elevated plasma levels found in areas with higher than optimal levels of fluoride in the water are also associated with fluorosis.

Commercial processing techniques can also introduce large quantities of fluoride into an infant's diet. The variability of the diet and the availability of fluoride from that diet, however, should both be considered. Children fed only breast milk in areas with optimally fluoridated water receive low doses of fluoride from that milk. Limited exposure to fluoride supplements, however, may have only limited caries preventive benefits for these children.

During pregnancy, the human placenta is permeable to fluoride and concentrations of fluoride in fetal bones and teeth increase with the age of the fetus. Fetal blood fluoride concentrations approximate maternal levels at the steady state. Periodic and abrupt elevations in the fluoride concentrations of the maternal blood, however, are not detected in the same amounts in the fetal blood. The results of clinical studies determining the effect of prenatal fluorides remain equivocal.

Significant reductions in the prevalence of dental caries occur when optimal concentrations of fluoride are introduced into drinking water.[1] Millions of people in the United States and other countries, however, do not have access to optimally fluoridated water due to lack of central water supplies or the decision not to have fluoride in their drinking water. Multiple clinical trials have shown that postnatal fluoride supplements can provide an alternate method of reducing caries to children living in suboptimally fluoridated areas.[2] Prenatal fluoride supplements have also been suggested as a means of providing the earliest possible caries prevention benefits.[3–6]

The purpose of this paper is to review and discuss the following:

(1) Commercially available fluoride supplements and the current supplemental fluoride schedule;

(2) The need for fluoride and vitamin supplementation and the topical effects of supplements;

(3) Supplementation during breast-feeding and total dietary fluoride;

(4) The rationale for prenatal fluoride supplements;

(5) The efficacy of postnatal fluoride supplements.

COMMERCIALLY AVAILABLE PRODUCTS

A variety of fluoride (F) supplements are commercially available (Tables 5–1 and 5–2). Fluoride supplements in liquid form are preferred for infants, young children, and individuals who have problems handling tablets. Liquid supplements in drop form are available without vitamins in the following doses: 0.125 mg F/drop, 0.25 mg F/drop, and 0.5 mg F/ml; whereas liquid supplements with vitamins are available in the following doses: 0.25 mg F/ml, 0.5 mg F/0.6 ml, and 0.5 mg F/ml. Supplements in tablet form for the older patient are available without vitamins in the following doses: 0.25 mg F, 0.5 mg F, or 1 mg F. The 1-mg F tablet can be obtained as neutral sodium fluoride (NaF) or as acidulated phosphate fluoride (APF). Tablets with vitamins are only available in 0.5 mg F or 1 mg F doses. APF mouthrinses with a pH of 4.0 have been formulated so that a 5-ml quantity can be swished in the mouth for topical effect, then swallowed, providing 1 mg F for systemic effect.

Ingredients in some supplements include flavorings, coloring agents, and small quantities of alcohol. Sucrose, dextrose, mannitol, glycerin, and saccharin are added in various amounts to pediatric vitamin fluoride preparations (e.g., Vi-Daylin tablets contain 250 mg sucrose/tablet; Poly-Vi-Flor tablets, 225 mg sucrose/tablet and 225 mg dextrose/tablet; Vi-Penta-F multivitamin drops, 687.5 mg glycerin/ml and 3 mg saccharin/ml).[7] Various preparations are available without sugar, saccharin, or coloring agents for children whose parents request that type of formulation. The necessity of flavoring agents in fluoride supplements is questionable since salivary stimulation by flavored tablets has been implicated in lower salivary fluoride levels.[8,9] The effect on cariostasis, however, is unknown.

All vitamin fluoride preparations contain vitamins A, D, and C, although the amounts may vary between products. Some vitamin fluoride supplements also contain mixtures of E, B_1, B_3, B_5, B_6, B_{12}, and/or iron.

The multiplicity of formulations is confusing when writing a generic prescription. Generic prescriptions are frequently written incorrectly because clinicians do not know the equivalent amount of sodium fluoride necessary to give the desired amount of fluoride. Prescription pads produced by a drug company are frequently utilized to facilitate prescription writing.

Cost should be considered when prescribing a supplement. Sodium fluoride tablets of varying dosages (i.e., 0.25, 0.5, and 1 mg F/tablet) from the same company cost essentially the same. Therefore, when a child requires a higher fluoride dosage due to age, a new prescription for increased dosage per tablet rather than an increased number of tablets should be written. Otherwise, the cost to the parent is approximately doubled.

The choice between APF tablets and neutral sodium fluoride tablets can be based

Table 5–1. Fluoride Supplements Available in Liquid Form With or Without Vitamins

Form	Dose	Brand Name	Company	Quantity
Drops	*0.125 mg F/Drop (.275 mg NaF/Drop)	Karidium#	Lorvic	30/60 ml
		Luride#	Hoyt	30 ml
	0.25 mg F/Drop (0.55 mg NaF/Drop)	Fluoritab	Fluoritab	19 ml
		Flura Drops#	Kirkman	24 ml
		Pediaflor	Ross	50 ml
	0.5 mg F/ml (1.1 mg NaF/ml)	Pediaflor	Ross	50 ml
Drops with Vitamins	0.25 mg F/ml (0.55 mg NaF/ml)	Abdec with Fluoride Baby Vitamin	Parke-Davis	50 ml
		Poly-Vit Fluoride	Rugby	50 ml
		Poly-Vi-Flor 0.25 mg.	Mead Johnson	50 ml
		Tri-Vi-Flor 0.25 mg.	Mead Johnson	50 ml
		Tri-Vi-Flor 0.25 with Iron	Mead Johnson	50 ml
		Vi-Daylin F⁰	Ross	50 ml
		Vi-Daylin F† Iron	Ross	50 ml
		Vi-Daylin/F ADC	Ross	50 ml
		Vi-Daylin/F ADC + Iron	Ross	50 ml
	0.5 mg F/0.6 ml (1.1 mg NaF/0.6 ml)	Adeflor	Upjohn	30 ml/150 ml
		Vi-Penta F Infant	Roche	30 ml
		Vi-Penta F Multivitamin	Roche	30 ml
	0.5 mg/ml (1.1 mg NaF/ml)	Dentavite	Reid-Provident	50 ml
		Florvite Pediatric	Everett Labs	50 ml
		Florvite + Iron Pediatric	Everett Labs	50 ml
		Ped-Vite with Fluoride	Three P	50 ml
		Polysorbin F	Reid-Provident	50 ml
		Poly-Vi-Flor 0.5 mg	Mead Johnson	30 ml/50 ml
		Poly-Vi-Flor with Iron	Mead Johnson	50 ml
		Polyvite with Fluoride	Geneva Generics	50 ml
		Tri-Bay-Flor	Bay	50 ml
		Trisorbin F	Reid-Provident	50 ml
		Tri-Vi-Flor 0.5 mg	Mead Johnson	30 ml/50 ml
Rinse	0.2% F 1 mg F/5 ml APF	Phos-Flur#	Hoyt	250/500 ml
		NaF Rinse Acidulated	Orachem	500 ml
		Oral Rinse and Systemic#	Pharm.	

*Ionic fluoride dose is given first. Dosage of fluoride compound appears in parens.
⁰ - Sugar and saccharin-free
† - Dye-free
# - ADA approved

Based on information from:
(1) Facts and Comparisons
(2) Physicians Desk Reference
(3) Accepted Dental Therapeutics

on availability at the pharmacy. Although manufacturer's prices for APF tablets were originally more than twice those for the neutral fluoride tablets, manufacturer's current prices for APF tablets are somewhat lower than for neutral fluoride tablets. The cariostatic effects of APF tablets appear to be equivalent to neutral fluoride tablets,[2]

Table 5–2. Fluoride Supplements Available in Tablet Form With or Without Vitamins

Form	Dose	Brand Name	Company	Quantity
Tablet	*0.25 mg F (0.55 mg NaF)	Luride 0.25 Lozi-Tabs[0#]	Hoyt	120
	0.5 mg F (1.1 mg NaF)	Luride 0.5 Lozi-Tabs[0#]	Hoyt	120/1200
	1 mg F (2.2 mg NaF)	Fluorineed	Hanlon	100/1000
		Fluoritab[#]	Fluoritab	100
		Flura[†#]	Kirkman	100
		Flura-Loz[†#]	Kirkman	100
		Karidium[#]	Lorvic	180/1000
		Luride Lozi-Tabs[0#]	Hoyt	120/1000
		Luride-SF Lozi-Tabs[0#†]	Hoyt	120
	(1 mg APF)	Phos-Flur[#]	Hoyt	120/500
Tablets with Vitamins	0.5 mg F (1.1 mg NaF)	Adeflor Chewable 0.5 mg	Upjohn	100/500
		Caritab Softab	Stuart	100
		Poly-Vi-Flor Chewable 0.5	Mead Johnson	100
	1 mg F (2.2 mg NaF)	Adeflor Chewable 1 mg	Upjohn	100/500
		Dentavite Chewable	Reid-Provident	100
		Florvite	Everett	100/1000
		Flura-Vite with Fluoride	Kirkman	100
		Mulvidren-F Softab	Stuart	100
		Poly-Vi-Flor Chewable	Mead Johnson	100/1000
		Poly-Vi-Flor with Iron	Mead Johnson	100/1000
		Tri-Vi-Flor	Mead Johnson	100/1000
		Vi-Daylin Chewable	Ross	100
		Vi-Penta F Chewable	Roche	100
		Vita-Flor	Rugby	100/1000
		Vi-Daylin/F & Iron Chewable	Ross	100

*Ionic fluoride dose is given first. Dosage of fluoride compound appears in parens.
[0] - Sugar and saccharin-free
[†] - Dye-free
[#] - ADA approved

Based on information from:
(1) Facts and Comparisons
(2) Physicians Desk Reference
(3) Accepted Dental Therapeutics

although significantly decreased fluoride concentrations in the saliva occur following APF tablet administration as compared to fluoride concentrations following neutral fluoride tablet administration.[10] The significance of the lower salivary levels is unknown since clinical efficacy is similar.

Fluoride-Vitamin Supplements

Fluoride vitamin combinations are as effective as fluoride supplements alone.[11] If vitamins were indicated for a patient living in an area with suboptimal concentration of fluoride in the water, the combination supplement would facilitate administration and provide definite caries benefits.[2]

Although fluoride-vitamin supplements have been recommended to improve the problems with compliance associated with fluoride supplements,[12] there seems to be little basis for recommending these supplements to the average normal patient. In 1980, the Committee on Nutrition from the American Academy of Pediatrics[13] reviewed the vitamin and mineral supplement needs of children in the United States. There was insufficient evidence for routine vitamin and mineral supplementation in properly nourished normal children, full-term, formula-fed infants, or most breast-fed, full-term infants. National dietary and health surveys[14–16] found little evidence of vitamin or mineral insufficiencies except for iron. Total insufficiency of food was considered the major nutritional problem in preschool children of lower socioeconomic status.

Multivitamin supplements were indicated, however, in some situations such as:

(1) Children and adolescents from deprived families who did not have adequate diets or who suffered from parental neglect or abuse;

(2) Children and adolescents with anorexia, poor and capricious appetites, or poor eating habits;

(3) Children on dietary regimens to manage obesity;

(4) Children and adolescents consuming vegetarian diets without adequate dairy products;

(5) Documented cases of dietary insufficiencies in breast-fed infants of a malnourished mother.

Topical Effects

Significant topical benefits can be obtained from supplemental rinses or chewable fluoride tablets. Caries reductions of 20 to 80% for erupting permanent teeth are reported in school programs where children are instructed to chew or suck the tablets prior to ingestion.[17–19] This is in contrast to a zero caries reduction in a study in which children were told to swallow the pills directly.[20] Topical effects are also noted for primary teeth.[2]

Patients should be instructed to dissolve or chew the tablet rather than swallow the tablet directly. Rinse supplements should be vigorously swished around the mouth with the liquid forced interproximally. Only slight increases (less than 0.05 ppm F) in salivary fluoride concentrations occur when a 1-mg F supplement is swallowed without chewing or dissolving.[21,22] In contrast, when tablets are chewed, then swished and swallowed or allowed to dissolve slowly, fluoride concentrations in the saliva are significantly elevated (12 to 0.4 ppm F) for at least 1 hour.[10,23] Salivary fluoride levels are higher when tablets are allowed to dissolve slowly compared to when they are sucked.[8] Patients should be told to move a slowly dissolving tablet throughout the mouth, since fluoride from this source is most concentrated in the area of the tablet.[23]

The mode of topical action by supplements is not clear. Increase in the fluoride concentration in the outer layer of erupted enamel has been suggested as a cariostatic mechanism in erupted teeth. However, studies show conflicting results. Mellberg et al.[24] found statistically significant differences in the fluoride content of the outer 5-μm layer of enamel in children receiving sodium fluoride tablets during a 3-year school program as compared to controls. These findings were not confirmed by Shern et al.[25] in school children who received APF tablets for a 3-year period. Significant differences in fluoride concentrations in the outer layer of enamel of children 7 to 12 years of age who ingested supplements from birth are reported.[26] The amount of post-eruptive uptake of fluoride is not clear. However, in a study that lasted approximately 17 months following discontinuation of tablets, Bruun et al.[27] did not find statistically significant differences in the outer layer of the enamel of premolars in children taking fluoride supplements as compared to controls. The crowns of the premolars had been formed but were unerupted for 3 years during the study. The tablets were discontinued either prior to or just as the teeth were erupting.

Fluoride supplementation has not been shown to significantly affect the colonization of *Streptococcus mutans* on the smooth surfaces of teeth in human subjects who ingested fluoride supplements from birth.[28] However, significantly higher percentages of *S. mutans* were found in first permanent molar fissures of patients who ingested supplements as compared to controls.[28]

The caries preventive effects on permanent teeth that were erupting when supplementation began appear to be lost following cessation of fluoride supplements.[29] Teeth that lay in the follicle for years during supplementation maintain significant caries preventive effects longer than those erupting when supplementation started.[26,29] Further studies on the topical effects of fluoride supplementation and means of prolonging these effects are necessary.

CURRENT SUPPLEMENTAL FLUORIDE SCHEDULE

The current supplemental fluoride dosage schedule (Table 5–3) was established in 1979.[30] Dosage is dependent on age of the child and concentration of fluoride in the child's drinking water. Estimates of total daily fluoride intake of children 1 to 12 years old[31] living in a fluoridated area were used as a basis for determining daily dosage.

In areas with drinking water containing less than 0.3 ppm F, children aged 2 weeks to 2 years should receive 0.25 mg F/day, whereas children aged 2 to 3 years should receive 0.5 mg F/day, and children 3 to 13 years should receive 1.0 mg F/day. If the drinking water contains 0.3 to 0.7 ppm F, then children below the age of 2 years should receive no supplements, children aged 2 to 3 years should receive 0.25 mg F/day, and children over 3 years, 0.5 mg F/day. No supplements should be prescribed for children living in areas with greater than 0.7 ppm F in the drinking water.

Usually, the local health department or water department can provide information concerning the fluoride concentration of a patient's drinking water. If it is unknown, the clinician must have a sample of the drinking water analyzed for fluoride concentration prior to prescribing supplements.

Fluorosis, characterized by opaque enamel areas varying in size depending on the severity of the condition, can occur due to incorrectly prescribing too great a dose of fluoride supplements. Although mild forms of fluorosis cannot be observed by

Table 5–3. Dosage Schedule for Fluoride Supplements

Age (Years)	Concentration of Fluoride in Water (ppm)		
	<0.3	0.3–0.7	>0.7
0–2	0.25 mgF/day	0.0 mgF/day	0.0 mgF/day
2–3	0.50	0.25	0.0
3–13	1.00	0.50	0.0

most people, the moderate and severe forms resulting in staining and pitting of the enamel are esthetically undesirable.

Maximum cariostatic benefits to both primary and permanent teeth are obtained when fluoride supplements are administered as soon as possible after birth.[2] Significant benefits to permanent teeth, however, are still obtained if supplementation is delayed.[2]

Total Fluoride

The fluoride concentrations in human breast milk and cow's milk are generally reported as less than 0.05 mg/liter.[32–35] Therefore, infants who were exclusively breast-fed or fed cow's milk even though they lived in fluoridated areas should be given fluoride supplements.[34] Caution should be applied, however, in prescribing supplements to these children. Introduction of additional baby foods at early ages is well documented[36–40] with as few as 10% of infants breast-fed for as long as 4 months.[37] Frequent contact with the mother to ascertain changes in feeding habits and strict instructions concerning discontinuation of supplements if additional foods or fluoridated water are added to the diet should be considered. The efficacy of fluoride supplements on breast-fed children living in fluoridated areas who ingest fluoride supplements for only a few months still needs to be determined, although studies[41] suggest that usage of supplements for short periods gives limited caries reductions.

Ingestion of fluoride by the mother does not appreciably elevate the fluoride concentration in the breast milk[42,43] and is not recommended as a means of supplying additional fluoride to the breast-fed child.

Commercial processing of baby foods has generated concerns[34,44–46] that infants may receive more than the optimal amount of fluoride, which was estimated to be 0.5 mg F/day by McClure[31] or 0.05 to 0.06 mg F/kg by other researchers.[2,44]

The fluoride concentrations in dry infant cereals, fruit juices, and milk formulations are increased when processed with fluoridated water as compared to those processed with nonfluoridated water.[34,46] Foods containing chicken are found to have higher and more variable fluoride contents than other meat products due to inclusion of bony fragments. The fluoride content of other meat, vegetable, and fruit products appears to be affected to a lesser extent by commercial processing.[46]

Calculations by various researchers concerning maximum and minimum fluoride intakes for infants and children vary depending on the methodology of fluoride analysis and the specific kinds and amounts of each foodstuff in the diets. Although Wiatrowski et al.[45] estimated considerably greater maximum fluoride intakes for infants than considered optimal, the study is generally disregarded today due to the analytic method of fluoride analysis, which gave falsely high values to the diet they constructed. Other investigators, reporting much lower total daily fluoride intakes, calculated maximum intakes that exceeded the optimal values for infants.[34] A particular problem was identified with infants in fluoridated areas who were fed concentrated formulas that were high in fluoride and diluted with fluoridated water.[34] These children potentially received 2 to 3 times the optimal levels of fluoride. Another area of concern was the high fluoride concentration of ready-to-feed formulas.[34] These formulas could potentially be given to infants living in suboptimally fluoridated areas and receiving fluoride supplements. As a result of these studies, manufacturers in the United States have reportedly removed major amounts of fluoride from their processing techniques. Additional studies of the diets of 6-month old infants indicate that although the total fluoride content of the diets vary considerably even between fluoridated areas, total daily fluoride intake does not exceed optimal levels.[47]

Calculations of maximum and minimum fluoride values, however, have limiting features. Not only are infant diets highly variable, but the amount and rate of fluoride absorption from food varies. Decreased absorption rates and plasma fluoride levels are documented in individuals who take fluoride with milk.[35,43] Ericsson[35] suggested that fluoride absorption from fluoride supplemented milk was complete although delayed; however, other studies[43] indicated that fluoride absorption from supplemented milk was incomplete and that the ionic fluoride in milk may be partially bound.[32,48]

Fluoride from solids is less available than fluoride from liquids. The availability of fluoride from foodstuffs depends on the following: the presence of food, which may act as a physical barrier to diffusion of fluoride to the gastrointestinal mucosa; the complete breakdown and release of fluoride bound to various foods; and the presence of other interfering ions, such as calcium, magnesium, and aluminum.

Only 55% of the fluoride available in Pablum, a baby cereal containing large amounts of bone, was absorbed.[47,49] An individual fed 2.5 mg F as strained chicken containing bony fragments showed delayed and lower plasma fluoride concentrations[50] as compared to one ingesting 2.5 mg F in NaF with water (Fig. 5–1). The same in-

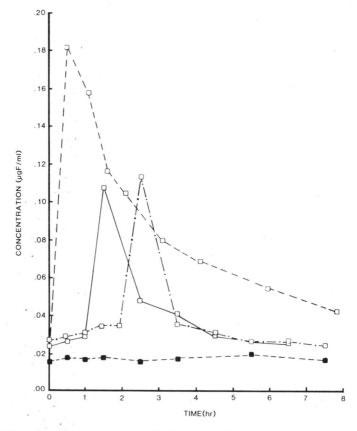

Fig. 5–1. Plasma fluoride concentrations following 2.5 mg F as NaF (- -), rice cereal (·-·) and strained chicken (—) as compared to baseline concentrations (■- -)

dividual also demonstrated delayed and decreased plasma fluoride levels following ingestion of cereal.[50] Low plasma availability of fluoride from bone meal and cereal has also been reported in animals.[51–53] Foods such as cereals and strained chicken contain high concentrations of fluoride, which appears to be bound (i.e., to bone) but is released in an acidic environment. Methods of fluoride analysis utilizing acid diffusion to release fluoride require hours to allow complete release of fluoride present in food.

Solid foods are regarded as making only a small contribution to the total daily intake of fluoride.[31,47] Major contributors are water and other fluids such as formula or juices.

Influence of Diet on Fluorosis. Plasma fluoride concentrations, which are implicated as factors in fluorosis,[54] are related to dietary substances in liquid or solid form. The relationship between fluorosis and diet should be clinically investigated. Diets containing large quantities of fish bone have been implicated in fluorosis.[55] However, no significant differences in fluorosis of permanent teeth could be found in children who were fed formula diluted with tap water containing 1.2 ppm F as compared to children who were breast-fed.[55a] Few reports associate fluorosis in humans with food intake other than water.[55b] Further study is required to determine the effect of the fluoride content of various foods on fluorosis.

Body weight is also implicated in fluorosis. A daily dose of 0.1 to 0.3 mg F/kg body weight can cause enamel fluorosis in cattle[56] and rats.[57] Angmar-Manson and Whitford[54] found that a total daily dose of 0.13 mg F/kg was not the important consideration in fluorosis, but that 0.13 mg F/kg was the amount given at one time. Steady state levels of 3.3 μM F were also associated with fluorosis.[54]

If the fluorosis-causing dose of 0.13 mg F/kg can be applied to humans, the dose of 0.25 mg F/day given as a single bolus may produce fluorosis in a low birth weight child of approximately 4 lb 3 oz. If one looks at a former supplemental schedule that recommended 0.5 mg F/day for children 0–2 years old, babies below the approximate weight of 8 lb 8 oz are potentially susceptible to fluorosis. Using the dose of 0.1 mg F/kg as the borderline dose for fluorosis, children below 11 lb are susceptible to fluorosis when given a 0.5-mg fluoride supplement.

The schedule for infants to 2-year olds was reduced to 0.25 mg F/day from 0.5 mg F/day due to reports of fluorosis. Aasenden and Peebles[26] reported the occurrence of fluorosis in 84% of their subjects who ingested 0.5 mg F from birth until 3 years and 1.0 mg F thereafter. Although most cases of fluorosis were considered questionable to mild in severity, 14% of the patients exhibited fluorosis of moderate severity. However, even the dose of 0.25 mg F/day may be a borderline dose since mild forms of fluorosis are reported[58] when 0.25 mg F/day is given before 24 months of age.

PRENATAL FLUORIDES

In 1966, the Food and Drug Administration banned the advertisement of fluoride products that claimed to cause significant caries reductions in the offspring of pregnant women. Safety was not a factor in this decision. Lack of adequate documentation concerning clinical efficacy was emphasized. Efficacy of prenatal fluorides is still a controversial topic.[59–61]

A discussion of the rationale for the use of prenatal fluorides in preventing dental caries may include several topics, such as those on the following page:

(1) Transfer of fluoride from the maternal circulation to the fetal circulation;
(2) Clinical trials with prenatal fluorides.

Placental Transfer of Fluoride.   The permeability of the placenta to fluoride has been questioned due to several observations, including:

(1) Paucity of published articles reporting fluorosis in deciduous teeth;
(2) High concentration of fluoride in the placenta;
(3) Differences in fluoride concentrations in hard tissues of fetuses formed in areas with different concentrations of fluoride in the water;
(4) Low fluoride concentrations in fetal blood following a fluoride challenge to the maternal circulatory system.

Fluorosis in Primary Teeth.   Controversy concerning the permeability of the placenta to fluoride stems from early conflicting reports[62,63] concerning the occurrence of fluorosis in primary teeth. Since primary teeth calcify primarily before birth and few reports existed of fluorotic primary teeth, fluoride was assumed to have limited access to the fetus.

Thylstrup[41] found that the primary teeth of all children in areas where water contained 3.5, 6.0, or 21.0 ppm F exhibited varying degrees of fluorosis. Although the severity of macroscopic changes was generally less pronounced than in the corresponding permanent teeth, the degree of severity was positively associated with the concentration of fluoride in the water supplies. Thylstrup suggested that the appearance of fluorosis in primary teeth may be affected by enamel thickness, and that no human placental barrier exists to fluoride.

Prenatal Fluoride Metabolism.   Metabolism of a prenatal fluoride supplement is complex and is not fully understood. However, a brief review of current knowledge concerning fluoride metabolism may increase the understanding of prenatal fluoride supplementation. A fluoride supplement is absorbed primarily in the maternal stomach and, to a lesser degree, in the small intestines[64,65] (Fig. 5–2). Fluoride, hypothetically, diffuses across the gastric mucosa as hydrogen fluoride,[66] then dissociates in the circulatory system to yield fluoride ions. While plasma fluoride concentrations increase rapidly, diffusion into maternal extracellular and intracellular fluids, uptake by maternal bone, and excretion by the kidneys begin within a few minutes following ingestion. Plasma fluoride maxima are reached within 30 to 60 minutes,[35] after which time excretion of fluoride from the plasma exceeds absorption. Ultimately, most of the fluoride is either taken up by mineralized tissue or excreted in the urine.[67] Unabsorbed fluoride is lost in the feces.

Fluoride does diffuse across the placenta[68–74] (Fig. 5–3). Reports of fluoride concentrations in fetal blood vary from the same to less than 75% of maternal blood at the steady state.[68–70,75] These studies were accomplished on either pregnant women or sheep under anesthesia. One study,[75] which reported complete transfer of fluoride to the fetus, has been criticized for measuring total fluoride, not ionic fluoride, which is more readily exchangeable.

Early investigators[75] reported high fluoride concentrations in the placenta and suggested that the placenta acted either as a barrier to prevent more than trace amounts of fluoride from reaching the fetus, or as a storage depot so that the fetus would have adequate fluoride for development. The high fluoride concentrations in the ma-

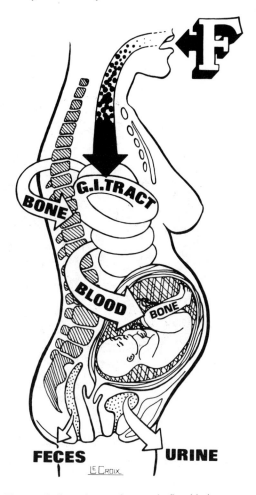

Fig. 5–2. The metabolic pathway of systemic fluoride in a pregnant woman.

ture placenta, however, may be due to calcific areas, which bind increasing amounts of organic fluoride with age.[68,77]

Fluoride is taken up by the fetal bones and teeth.[78–80] Analysis of human fetuses indicate that the fluoride content of bones and teeth increases with the age of the fetus and that the fluoride uptake is greater in bones than teeth.[78,81] Although significant increases in fluoride concentrations of these tissues were found in fetuses from areas with 0.5 to 0.6 ppm F water as compared to those from areas with 0.05 to 0.1 ppm F water, Gedalia et al.[79] suggest that fluoride metabolism from drinking water may be limited since there were no significant differences between tissues formed at 1 ppm F and those formed at 0.5 to 0.6 ppm F.

At approximately 14 weeks in utero, the kidneys start to function and excrete increasing amounts of fluid into the amniotic fluid.[82] By term, approximately 500 ml of amniotic fluid is swallowed daily by the fetus and probably reabsorbed into the fetal GI tract.[83] It may be removed through the placenta into the mother's circulatory system or excreted again. The significance of these processes in fluoride metabolism, however, is unknown.

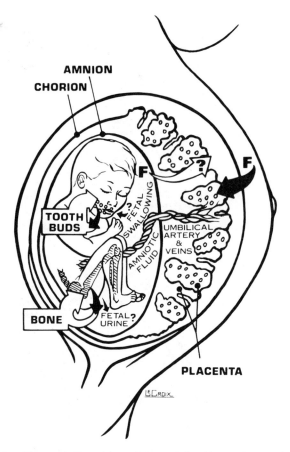

Fig. 5–3.   The metabolic pathway of prenatal fluoride supplements in a fetus.

Elevated Maternal Blood Concentrations.   The rate of transfer of fluoride to the fetus during sudden elevations in maternal blood concentrations is controversial. Plasma fluoride concentrations in humans drinking suboptimally fluoridated water rise from baseline concentrations of approximately 0.01 ppm F to maxima of less than 0.07 ppm F within 30 minutes after ingestion of 1.5 mg F.[84] Plasma fluoride concentrations decrease rapidly and reach baseline concentrations within 7 hours. However, lower plasma fluoride concentrations and a shorter period of elevated plasma concentrations would be expected following ingestion of a fluoride supplement that contains only 1 mg fluoride. If the supplement is taken with food, milk, or calcium-rich pills, decreased concentrations of fluoride in the plasma should be expected.[85]

Studies with pregnant women, sheep, and rabbits treated with an intravenous bolus of fluoride show rapid maternal clearance with slight increases in fetal blood fluoride,[71,86] although one study demonstrated fetal serum values as high as maternal values in some sheep.[70] Generally, fluoride concentrations in the fetal blood do not exceed 25% of the fluoride concentrations of maternal blood although slight elevations are demonstrated.

Studies by Feltman and Kosel[87] suggest that fluoride concentrations of fetal blood are elevated when mothers either receive a 1-mg F tablet or drink artificially fluoridated water during pregnancy as compared to fetal cord blood when mothers drank either nonfluoridated water or received no supplement. However, comparison between the two groups of controls is difficult since the concentrations of fluoride in their drinking water is not reported. Maternal fluoride concentrations were also not reported.

Prenatal Exposure to Fluoridated Water

The effect on dental caries of prenatal exposure to optimally fluoridated water was reported in eight studies. Conclusions based on the composite of the studies are difficult. Blayney and Hill[88] reported caries reductions for primary teeth of 30 to 35% for children exposed to fluoridated water pre- and postnatally. Although the caries reductions for children exposed both pre- and postnatally are claimed to be significant, it is not clear whether a statistical analysis was done. Tank and Storvick[89] also compared caries reductions of controls to children exposed only postnatally and to children exposed pre- and postnatally to fluorides. Although they found a greater caries reduction in the pre- and postnatally exposed group as compared to the group exposed only postnatally, the difference was not significant. Since their results were consistent with studies of fluoride analysis of fetuses, they concluded that prenatal exposure to fluoride produced pronounced reductions in dental caries rates.

Horowitz and Heifitz[90] subdivided a population of over 2,500 children into five groups based on the length of their prenatal exposure to fluoridated water. They found very low inverse relationships between caries experience in deciduous teeth and prenatal exposure to fluoridated water. Small differences in caries prevalence for permanent molars were found between groups. They concluded that the relationship, if any, between prenatal exposure to fluoride and caries prevalence in deciduous cuspids, molars, and permanent first molars is minor and has no practical significance in the prevention of decay.

Katz and Muhler[91] also found small and statistically insignificant decreases in caries prevalence in deciduous teeth of children when their mothers were exposed prenatally to increased amounts of fluoridated water. They indicated that larger studies should be conducted to clarify their results, since first premanent molars did show significant caries reductions when the children were exposed to prenatal fluorides.

In contrast, Carlos et al.[92] found minor differences in the prevalence of dental caries in primary teeth or caries-free children between the control group and groups of children who either were exposed to pre- and postnatal fluoride or to postnatal fluoride only.

Following discontinuation of fluoridation in Antigo, Wisconsin, the increase in dental caries was so great that Lemke et al.[93] concluded that fluoridated water prenatally was ineffective in preventing dental caries. Arnold et al.[94] reported greater caries reductions in children with both prenatal and postnatal exposure to fluoridated water than those children who were exposed to fluoridated water postnatally. Lewis et al.[95] reported that little differences existed between the treatment needs of children who received fluoridated water prenatally and postnatally compared to those receiving fluoridated water postnatally only.

Clinical Studies with Prenatal Fluorides

Few studies have evaluated the efficacy of prenatal fluoride supplements clinically. Although the studies all suggest a positive effect on caries reductions, problems associated with study designs, results, and conclusions detract from their credibility.

Feltman and Kosel[96] studied three groups of children for 14 years: one receiving prenatal fluoride supplements only, one receiving prenatal and postnatal supplements for 3 to 8 years, and one group received no supplementation. Caries reductions were greatest for the group receiving both pre- and postnatal supplements, although pronounced reductions were found in children with prenatal supplementation only. Doubts concerning retention of prenatal fluoride benefits for years following discontinuation have been raised. Failure to randomly assign subjects to study groups has also been criticized.

Hoskova[97] reported definite caries reductions in primary teeth of children receiving pre- and postnatal fluoride supplements compared with a control group who received no fluoride supplementation or a group who received postnatal supplements in nursery and kindergarten programs. The investigator attributed the greater caries reduction in the prenatal groups to parent compliance with daily administration at home versus administration on school days, and lack of administration of postnatal supplements for 2 years during the middle of the study.

Kailis et al.[98] reported significant reductions (82%) in primary teeth of children 4 to 6 years of age who received prenatal and postnatal supplements from birth compared to those children who received postnatal supplements from birth and children who received no supplements. Although well-designed and reported, the study was retrospective, thus relying on the memory of parents for a 5 to 7 year period.

Pritchard[99] also reported greater caries reductions in primary teeth of children exposed to prenatal and postnatal supplements as compared to postnatal use only or to controls with no supplements. The study has been criticized from several points: lack of information concerning methodology, clarity, errors in tabulation, and incorrect statements in the text.

Schutzmannsky[100] reported significant caries reductions of 13% in primary canines and molars of children receiving only prenatal fluorides, 14% in children receiving postnatal supplements only, and 30% in children receiving both prenatal and postnatal fluorides. Permanent teeth had caries reductions of 6, 39, and 43%, respectively. Criticisms concerning lack of random assignment to study groups, long-term retention of cariostatic benefits from prenatal supplements only, lack of information concerning statistical tests, and other problems have been made.

A series of papers by Glenn[3-5] have reported definite caries reductions attributed to the use of prenatal fluoride supplements in optimally fluoridated areas. Serious problems are noted with the size of the populations (i.e., one patient only receiving prenatal fluorides in one of the studies), variables (i.e., patients receiving prenatal fluorides also received postnatal fluorides, fluoridated water and topical fluoride treatments), differences in ages of patients between study groups, lack of random assignment, and numerous other areas. Serious criticisms are raised concerning the methodology and validity of the conclusions.

Thus, there is a limited amount of clinical evidence to show that prenatal fluoride supplements do have cariostatic benefits. The safety of the fetus and mother is not questioned. Prenatal fluoride supplementation particularly in fluoridated areas cannot be recommended for public health measures due to the numerous problems associated with the clinical studies that have been reported. A large clinical study is now being conducted under the auspices of the National Institutes of Dental Research to determine the efficacy of prenatal fluorides.

REFERENCES

1. DHEW Bulletin No. FL-109: Evaluatory surveys of long-term fluoridation show improved dental health, 1979.
2. Driscoll, W.: The use of fluoride tablets for the prevention of dental caries. *In* International Workshop on Fluorides and Dental Caries Reductions. Edited by D. Forrester, and E. Schulz. Baltimore, Maryland, University of Maryland, 1974, pp. 25–96.
3. Glenn, F.B.: Immunity conveyed by a fluoride supplement during pregnancy. J. Dent. Child., *44*:391–395, 1977.
4. Glenn, F.B.: Immunity conveyed by sodium-fluoride supplement during pregnancy. Part II. J. Dent. Child., *46*:17–24, 1979.
5. Glenn, F.B., Glenn, W.D., and Duncan, R.C.: Fluoride tablet supplementation during pregnancy for caries immunity: A study of the offspring produced. Am. J. Obstet. Gynecol., *143*:560–564, 1982.
6. Glenn, F.B., Glenn, W.D., and Duncan, R.C.: Prenatal fluoride tablet supplementation and improved molar occlusal morphology: Part V. J. Dent. Child., *51*:19–23, 1984.
7. Schneiweiss, F.: Sweetener content of vitamin preparations. Am. J. Hosp. Pharm., *37*:1048, 1980.
8. McCall, D., Stephen, K.W., and McNee, S.G.: Fluoride tablets and salivary fluoride levels. Caries Res., *156*:98–102, 1981.
9. Bruun, C., and Givskov, H.: Fluoride concentrations in saliva in relation to chewing of various supplementary fluoride preparations. J. Dent. Res., *87*:1–6, 1979.
10. Parkins, F.M.: Retention of fluoride with chewable tablets and a mouthrinse. J. Dent. Res., *51*:1346–1349, 1971.
11. Hennon, D.K., Stookey, G.K., and Muhler, J.C.: Prophylaxis of dental caries: Relative effectiveness of chewable fluoride preparations with and without vitamins. J. Pediatr., *80*:1018–1021, 1972.
12. Margolis, F.J., et al.: Fluoride: Ten-year prospective study of deciduous and permanent dentition. Am. J. Dis. Child., *129*:794–800, 1975.
13. American Academy of Pediatrics Committee on Nutrition: Vitamin and mineral supplement needs in normal children in the United States. Pediatrics, *66*:1015–1021, 1980.
14. Dietary Intake Findings, 1971–1974: National Health Survey. DHEW Publication No. (HRA) 77-1647. Hyattsville, Maryland, National Center Health, Series 11, No. 202, 1977.
15. Owen, G., et al.: A study of nutritional status of preschool children in the United States, 1968–1970. Pediatrics, *53*:597–641, 1974.
16. Ten-State Nutrition Survey, 1968–70: Highlights. DHEW Publication No. (HSM) 72-8134. Atlanta, Center for Disease Control, 1970.
17. Stephen, K.W., and Campbell, D.: Caries reduction and cost-benefit after 3 years of sucking fluoride tablets daily at school, a double-blind trial. Br. Dent. J., *144*:202–206, 1978.
18. DePaola, P.F., and Lax, M.: The caries-inhibiting effect of acidulated phosphate fluoride chewing tablets: A two-year double-blind study. J. Am. Dent. Assoc., *76*:554–557, 1968.
19. Driscoll, W.: Effect of acidulated phosphate fluoride chewable tablets in school children: Results after 55 months. J. Am. Dent. Assoc., *94*:537–543, 1977.
20. Bibby, B.G., Wilkens, E., and Witol, E.: A preliminary study of the effects of fluoride lozenges and pills on dental caries. Oral Surg., *8*:213–216, 1955.
21. Shannon, I.L., and Edmonds, F.J.: Effect of fluoride dosage in human parotid saliva fluoride levels. Arch. Oral Biol., *17*:1303–1309, 1972.
22. Grøn, P., et al.: The direct determination of fluoride in human saliva by a fluoride electrode. Fluoride levels in parotid saliva after ingestion of single doses of sodium fluoride. Arch. Oral Biol., *13*:203–213, 1968.
23. Shern, R., Kennedy, J., and Bowen, W.H.: Fluoride levels of oral fluids from individuals using three regimens. Annual meeting of the USPH Professional Society, 1981, p. 14.
24. Mellberg, J.R., Nicholson, C.R., and Law, F.E.: Fluoride concentrations in deciduous tooth enamel of children chewing sodium fluoride tablets. J. Dent. Res., *51*:551–554, 1971.
25. Shern, R., Driscoll, W., and Korts, D.: Enamel biopsy results of children receiving fluoride tablets. J. Am. Dent. Assoc., *95*:310–314, 1977.
26. Aasenden, R., and Peebles, T.C.: Effects of fluoride supplementation on deciduous and permanent teeth. Arch. Oral Biol., *19*:321–326, 1974.

27. Bruun, C., et al.: Pre-eruptive acquisition of fluoride by surface enamel of permanent teeth after daily use of fluoride supplements. Caries Res., *17*:89–91, 1983.
28. Van Houte, J., Aasenden, R., and Peebles, T.C.: Oral colonization of *Streptococcus mutans* in human subjects with low caries experience given fluoride supplements from birth. Arch. Oral Biol., *23*:361–366, 1978.
29. Driscoll, W.S., Heifitz, S.B., and Brunelle, J.A.: Caries-preventive effects of fluoride tablets in school children four years after discontinuation of treatments. J. Am Dent. Assoc., *103*:878–881, 1981.
30. American Dental Association: *In* Accepted Dental Therapeutics. 37th Ed. Chicago, ADA, 1979.
31. McClure, F.J.: Ingestion of fluorine and dental caries. Quantitative relations based on food and water requirements of children to 12 years old. Am. J. Dis. Child., *66*:362–369, 1943.
32. Backer-Dirks, O., et al.: Total and free ionic fluoride in human and cows milk as determined by gas-liquid chromatography and the fluoride electrode. Caries Res., *8*:181–184, 1976.
33. Spak, C.J., Hardell, L.I., and DeChateau, P.: Fluoride in human milk. Acta Paediatr. Scand., *72*:699–701, 1983.
34. Adair, S.M., and Wei, S.H.: Supplemental fluoride recommendations for infant based on dietary fluoride intake. Caries Res., *12*:76–82, 1978.
35. Ericsson, Y.: The state of fluorine in milk and its absorption and retention when administered in milk. Acta Odontol. Scand., *16*:51–77, 1958.
36. Epps, R., and Jolley, M.: Unsupervised early feedings of solids to infants. Med. An. Dis. Columbia., *32*:493–495, 1963.
37. Fomon, S.E.: What are infants fed in the United States? Pediatrics, *56*:350–354, 1975.
38. Harris, L., and Chan, J.: Infant feeding practices. Am. J. Dis. Child., *117*:483–492, 1969.
39. Paige, D.: Avoiding overnutrition in infants. *In* Year one: Nutrition Growth Health. Edited by M. Syruck. Columbus, Ohio, Ross Laboratories, 1974, pp. 27–33.
40. Purvis, G.: What nutrients do our infants really get? Infant Nutr., *8*:28–34, 1973.
41. Thylstrup, A.: Distribution of dental fluorosis in the primary dentition. Community Dent. Oral Epidemiol., *6*:329–337, 1978.
42. Ekstrand, J.: No evidence of transfer of fluoride from plasma to breast milk. Br. Med. J., *283*:761–762, 1981.
43. Spak, C.J., Ekstrand, J., and Zylberstein, O.: Bioavailability of fluoride added by baby formula and milk. Caries Res., *16*:249–256, 1982.
44. Farkas, C.S., and Farkas, E.J.: Potential effect of food processing on the fluoride content of infant foods. Sci. Total Environ., *2*:399–405, 1974.
45. Wiatrowski, E., et al.: Dietary fluoride intake of infants. Pediatrics, *55*:517–522, 1975.
46. Singer, L., and Ophaug, R.: Total fluoride intake of infants. Pediatrics, *63*:460–466, 1979.
47. Ophaug, R.M., Singer, L., and Harland, B.F.: Estimated fluoride intake of 6-month old infants in four dietary regions in the U.S. Am. J. Clin. Nutr., *33*:324–327, 1980.
48. Duff, E.J.: Total and ionic fluoride in milk. Caries Res., *15*:406–408, 1981.
49. Ham, M., and Smith, M.: Fluorine balance studies on three women. J. Nutr., *43*:225–232, 1954.
50. Kula, K.S., Wei, S.H., and Wefel, J.: Availability of fluoride from certain baby foods. J. Dent. Res. *59*:309, 1980.
51. Richards, A., Fejerskov, O., and Ekstrand, J.: Fluoride pharmacokinetics in the domestic pig. J. Dent. Res., *61*:1099–1102, 1982.
52. Stillings, B.R., et al.: Further studies on the availability of the fluoride in fish protein concentrate. J. Nutr., *103*:26–35, 1973.
53. Zipkin, I., Lucas, S.M., and Stillings, B.R.: Biological availability of the fluoride of fish protein concentrate in the rat. J. Nutr., *100*:293–299, 1969.
54. Angmar-Manson, B., and Whitford, G.M.: Enamel fluorosis related to plasma F levels in the rat. Caries Res., *18*:25–32, 1984.
55. Pu, M.Y., and Lilienthal, B.: Dental caries and mottled enamel among Formosan children. Arch. Oral Biol., *5*:125–136, 1961.
55a. Ericsson, Y., and Ribelius, V.: Wide variations of fluoride supply to infants and their effect. Caries Res., *5*:78–84, 1971.
55b. Myers, H.: Fluorides and Dental Fluorosis. Monogr. Oral Sci., *7*:1–74, 1978.
56. Suttie, J.W.: Nutritional aspects of fluoride toxicosis. J. Anim. Sci., *51*:759–766, 1980.
57. Kruger, B.J.: The effect of different levels of fluoride on the ultrastructure of ameloblasts in the rat. Arch. Oral Biol., *15*:109–114, 1970.
58. Holm, A.K., and Anderson, R.: Enamel mineralization disturbances in 12-year-old children with known early exposure to fluorides. Community Dent. Oral Epidemiol., *10*:335–339, 1982.
59. Driscoll, W.: A review of clinical research on the use of prenatal fluoride administration for prevention of dental caries. J. Dent. Child., *48*:109–117, 1981.
60. Glenn, F.B.: The rationale for the administration of NaF tablet supplement during pregnancy and postnatally in a private practice setting. J. Dent. Child., *48*:118–122, 1981.

61. Thylstrup, A.: Is there a biological rationale for prenatal fluoride administration? J. Dent. Child., *48*:103–108, 1981.
62. McKay, F.S., and Black, G.V.: An investigation of mottled teeth: an endemic developmental imperfection of enamel of the teeth heretofore unknown in the literature of dentistry (I). Dent. Cosmos., *58*:477–484, 1916.
63. Smith, M.C., and Smith, H.B.: Mottled enamel of deciduous teeth. Science, *81*:77, 1935.
64. Stookey, G., Crane, D., and Muhler, J.: Effect of molybdenum on fluoride absorption. Proc. Soc. Exp. Biol. Med., *109*:580–582, 1962.
65. Wagner, M.: Absorption of fluoride by the gastric mucosa of the rat. J. Dent. Res., *41*:667–671, 1962.
66. Whitford, G., Pashley, D., and Stringer, G.: Fluoride renal clearance: A pH dependent event. J. Physiol., *230*:527–532, 1976.
67. Gedalia, I., Brzezinski, A., and Bercovici, B.: Urinary fluoride levels in women during pregnancy and after delivery. J. Dent. Res., *38*:548–551, 1959.
68. Shen, Y.W., and Taves, D.: Fluoride concentrations in the human placenta and maternal and cord blood. Am. J. Obstet. Gynecol., *119*:205–207, 1974.
69. Weiss, V., and DeCarlini, C.: Placental transfer of fluoride during methoxyflurane anesthesia for cesarean section. Experientia, *31*:339–341, 1975.
70. Maduska, A.L., et al.: Placental transfer of intravenous fluoride in the pregnant ewe. Am. J. Obstet. Gynecol., *136*:84–86, 1980.
71. Ericsson, Y., and Malmnas, C.: Placental transfer of fluoride investigated with F^{18} in man and rabbit. Acta Obstet. Gynecol. Scand., *41*:144–157, 1962.
72. Bawden, J.W., Wolkoff, A.S., and Flowers, C.E., Jr.: F^{18} recovery from fetal lambs following intravenous injection into the ewe. J. Dent. Res., *44*:1010–1014, 1965.
73. Murray, M.M.: Maternal transference of fluoride. J. Physiol., *87*:388–393, 1936.
74. Knouff, R.A., et al.: Permeability of placenta to fluoride. J. Dent. Res., *15*:291–294, 1936.
75. Armstrong, W.D., Singer, L., Makowski, E.L.: Placental transfer of fluoride and calcium. Am. J. Obstet. Gynecol., *107*:432–434, 1970.
76. Gardner, D.E., et al.: The fluoride content of placental tissue as related to the fluoride content of drinking water. Science, *115*:208–209, 1952.
77. Ericsson, Y., and Ulberg, S.: Auto radiographic investigations of the distribution of F^{18} in mice and rats. Acta Odontol. Scand., *16*:363–365, 1958.
78. Gedalia, I.: The fluoride content of teeth and bones of human fetuses. Arch. Oral. Biol., *9*:331–340, 1964.
79. Gedalia, I., Zukerman, H., and Leventhal, H.: Fluoride content of teeth and bones of human fetuses: In areas with about 1 ppm of fluoride in drinking water. J. Am. Dent. Assoc., *71*:1121–1123, 1965.
80. Brzezinski, A., Bercovici, B., and Gedalia, I.: Fluorine in the human fetus. Obstet. Gynecol., *15*:329–321, 1960.
81. Gedalia, I., Placental transfer of fluoride in the human fetus at low and high F^- intake. J. Dent. Res., *43*:669–671, 1964.
82. Walsh, S.Z., Meyer, W.W., and Lind, J.: The Human Fetal and Neonatal Circulation: Function and Structure. Springfield, Charles C Thomas, 1974, p. 351.
83. Natelson, S., Scommegva, A., and Epstein, M.: Amniotic Fluid. New York, John Wiley & Sons, 1974, p. 386.
84. Ekstrand, J., et al.: Pharmacokinetics of fluoride in man after single and multiple oral doses. Eur. J. Clin. Pharmacol., *12*:311–317, 1979.
85. Ekstrand, J., and Ehrnebo, M.: Influence of milk products on fluoride availability in man. Eur. J. Clin. Pharmacol., *16*:211–215, 1979.
86. Bawden, J.W., Wolkoff, A.S., and Flowers, C.E.: Placental transfer of F^{18} in sheep. J. Dent. Res., *43*:678–683, 1964.
87. Feltman, R., and Kosel, G.: Prenatal ingestion of fluorides and their transfer to the fetus. Science, *9*:560–561, 1955.
88. Blayney, J.R., and Hill, I.N.: Evanston dental caries study. XXIV. Prenatal fluorides-value of waterborne fluorides during pregnancy. J. Am. Dent. Assoc., *69*:291–294, 1964.
89. Tank, G., and Storvick, C.A.: Caries experience of children one to six years old in two Oregon Communities (Corvallis and Albany). I. Effect of fluoride on caries experience and eruption of teeth. J. Am. Dent. Assoc., *69*:749–757, 1964.
90. Horowitz, H.S., and Heifetz, S.B.: Effects of prenatal exposure to fluoridation on dental caries. Public Health Rep., *82*:297–304, 1967.
91. Katz, S., and Muhler, J.C.: Prenatal and postnatal fluoride and dental caries experience in deciduous teeth. J. Am. Dent. Assoc., *76*:305–311, 1968.
92. Carlos, J.P., Gittelson, A.M., and Haddon, W.: Caries in deciduous teeth in relation to maternal ingestion of fluoride. Public Health Rep., *77*:658–660, 1962.

93. Lemke, C.W., Doherty, J.M., and Ara, J.C.: Controlled fluoridation the dental effects of discontinuation in Antigo, Wisconsin. J. Am. Dent. Assoc., *80*:782–786, 1970.
94. Arnold, F., et al.: Effect of fluoridated public water supplies on dental caries prevalence. Public Health Rep., *71*:652–658, 1956.
95. Lewis, D.W., et al.: Initial dental care time, cost, and treatment requirements under changing exposure of fluoride during tooth development. J. Can. Dent. Assoc., *4*:140–144, 1972.
96. Feltman, R., and Kosel, G.: Prenatal and postnatal ingestion of fluorides—fourteen years of investigation—final report. J. Dent. Med., *16*:190–198, 1961.
97. Hoskova, M.: Fluoride tablets in the prevention of tooth decay. Cesk. Pediatr., *23*:438–441, 1968.
98. Kailis, D.B., et al.: Observations of the effects of prenatal and postnatal fluoride on some Perth pre-school children. Med. J. Aust., *2*:1037–1040, 1968.
99. Pritchard, J.L.: The prenatal and postnatal effects of fluoride supplements on West Australian school children, aged 6 to 8, Perth, 1967. Aust. Dent. J., *14*:335–338, 1969.
100. Schutzmannsky, G.: Fluorine tablet application in pregnant females. Dtsch. Stomatol., *2*:122–129, 1971.

Chapter 6
FLUORIDE MOUTHRINSES

James P. Carlos

Self-applied fluorides in the form of dilute mouthrinses are a highly cost-effective method of caries prevention, especially when used in school-based programs.

Sodium fluoride mouthrinses have been extensively studied in controlled clinical trials. Whether used daily, weekly, or fortnightly, sodium fluoride rinses have been consistently effective in partially reducing caries incidence in school-based children living in nonfluoride areas. Evidence for their effectiveness in fluoride areas is sparse, but positive.

Though less studied in controlled trials, stannous fluoride rinses can also be effective against caries. Recently, research interest has focused on the antiplaque properties of tin ion and the possibility that stannous fluoride rinses might prove useful in the prevention of gingivitis. Research has also suggested that sequential rinses with different fluoride compounds will have enhanced efficacy. Further clinical research is needed, however, to determine the answers to these questions.

Some evidence exists to suggest that fluoride rinses will also be effective in preventing root-surface caries in adults. This hypothesis is currently undergoing clinical testing.

As vehicles for the delivery of fluorides (F), mouthrinses are a relatively recent development, having been widely used in this country only for the past 8 to 10 years. Fluoride mouthrinses, together with fluoridated drinking water, soluble tablets, and dentifrices, fall into the category of self-applied preventive methods. These methods have great practical appeal because they minimize the need for supervision by trained personnel, and they are easily delivered to large numbers of people inexpensively.

A variety of fluoride compounds have been tested as rinses in animals and clinical studies, but only sodium fluoride mouthrinses are widely used in preventive programs. This chapter will review some of the clinical data on the efficacy of mouthrinses for the prevention of caries, and also comment briefly on the possible mechanisms by which they exert their effects.

Results of experiments will in most cases be cited as they have been presented in the research literature, without detailed critiques. Research studies, especially those involving humans, are almost always plagued by methodologic problems, which lead to varying degrees of imprecision in the results. Since it cannot be safely assumed that every study cited is of equal quality, the data should be considered in the aggregate when making judgments about the value of particular preventive regimens.

THE RATIONALE FOR FLUORIDE MOUTHRINSES

An underlying assumption of early research was that the therapeutic goal of topical fluorides was to decrease the acid solubility of enamel by converting hydroxyapatite crystals to fluorapatite. This reaction would occur when developing enamel was exposed to fluoride preeruptively. Consequently, topical fluoride formulations were first tested as solutions and gels containing relatively high concentrations of fluoride in an attempt to induce fluoride uptake by enamel.

An often cited example is a study in which a gel containing 1.10% sodium fluoride was applied in fitted trays to the teeth of children every school day for 21 months.[1] Caries incidence was 75% to 80% lower than that found in a placebo-treated control group, and large increases in the fluoride content of the superficial layers of enamel were also observed. This occurred mainly in the form of labile calcium fluoride, rather than fluorapatite, as it gradually decreased after the treatments were discontinued. In fact, it has not been possible to demonstrate clear correlations between bound fluoride in sound enamel and caries prevalence in epidemiologic surveys of populations exposed to systemic fluorides or topical fluorides.[2] Evidently, some other mechanism or mechanisms are involved in the therapeutic effects of systemic and topical fluorides.

It was later shown that very low levels of fluoride in the oral fluids are associated with more concentrated levels in plaque,[3] that low concentrations of fluoride are sufficient to inhibit glycolysis and acid production by plaque microorganisms[4] and, finally, that repeated exposure to low concentrations of fluoride effectively promotes remineralization of very early, incipient caries lesions.[5] Experiments with rats confirmed that protection against caries could be achieved by exposure of the teeth, posteruptively, to very low concentrations of fluoride, and that the protection did not depend on the uptake of fluoride by sound enamel.[6] These findings helped to explain the excellent clinical results already observed with dilute fluoride mouthrinses in several studies in Scandinavia. This and similar experimental evidence suggested that optimum preventive results with fluoride would be achieved by frequent exposure of plaque to low concentrations of the fluoride ion.

It has been shown that rinsing with dilute solutions of fluoride results in rapid elevation of plaque fluoride concentrations, but that levels have essentially returned to normal within 24 hours.[7] From this finding one might assume that daily rinsing regimens would be superior to once-a-week rinsing. The additional benefits of daily rinsing, however, have not been striking in controlled clinical trials. This apparent contradiction is not completely understood; therefore, it is possible that the predominant mechanism of action of the fluoride ion is the promotion of remineralization of early caries lesions, in which event weekly exposure to fluoride may suffice.

In the case of stannous fluoride, additional cariostatic mechanisms may be operating. Experimental and clinical evidence shows that stannous fluoride at concentrations normally used has an antibacterial effect against some plaque microorganisms,[8] and that this compound may also inhibit plaque formation by reducing the free surface energy of enamel.[9]

Sodium Fluoride Mouthrinses

Sodium fluoride (NaF) mouthrinses are usually formulated at concentrations of either 0.2% NaF (900 ppm F) for weekly use, or 0.05% NaF (225 ppm F) for daily use. They have been tested in both neutral and acidified forms in a water vehicle.

These rinses are intended to be used by forcefully swishing 10 ml of the liquid around the mouth for 60 sec before expectorating it. Because the the rinses should not be swallowed, they are not recommended for preschool-aged children, and children in kindergarten should rinse with 5 ml of solution, half the usual amount prescribed.

The results of some of the more recent clinical trials with neutral sodium fluoride mouthrinses are summarized in Tables 6–1 and 6–2.[10–22] These studies were reported between 1971 and 1982. Most of the studies involved the supervised use of sodium fluoride rinses by children living in nonfluoridated areas; the exceptions are also indicated in the tables.

Several general observations can be made about these data. With a few exceptions, most of the trials involved children who were aged 8 to 12 years at the beginning of the trials, which were continued for at least 2 years. Both daily rinsing, with approximately 250 ppm F, and weekly rinses with 900 ppm F were tested. In most of the studies, a sufficient number of subjects was available for follow-up providing reasonable confidence in the observed caries scores. Although the actual average observed difference in caries increment between test and control subjects differed considerably, the percent difference in the increment in test groups compared to controls was remarkably consistent in that it rarely was less than 25 to 30%. This was also true when, in one trial, fluoride mouthrinses were tested in an optimally fluoridated community.[22]

It may also be observed from Tables 6–1 and 6–2 that, within the limits shown, the concentration of fluoride in the rinses was not related to cariostatic efficacy. There is in these data the suggestion of increased efficacy of daily as compared to

Table 6–1. Results of Clinical Trials of Mouthrinsing with Neutral Sodium Fluoride Solutions (1971–1977)

Study	F Concentration (ppm)	Rinsing Frequency	Age at Baseline	Number Completing Trial	Length of Trial (Months)	DMFS Caries Increment Control	Test	Δ	% Difference
Horowitz et al. (1971)[10]	900	1/week S	6	129	20	1.3	1.1	0.2	16
(Same trial)	900	1/week S	11	117	20	2.9	1.7	1.2	44
Brandt et al. (1972)[11]	900	2/week S	11–12	94	21	7.0	4.0	3.0	43
Moreira & Tumang (1972)[12]	450	3/week S	7	50	24	7.5	4.0	3.5	47
(Same trial)	450	1/week S	7	50	24	7.5	5.7	1.8	25
(Same trial)	450	1/2 week S	7	50	24	7.5	5.8	1.7	23
Aasenden et al. (1972)[13]	200	1/day S	8–11	114	36	12.3	9.0	3.3	27
Heifetz et al. (1973)[14]	3000	1/week S	10–12	126	24	7.5	4.7	2.8	38
Rugg-Gunn et al. (1973)[15]	225	1/day S	11–12	222	34	10.2	6.6	3.6	36
Gallagher et al. (1974)[16]	1800	1/week S	10–11	306	24	4.4	2.9	1.5	34
Maiwald & Padron (1977)[17]	900	1/2 week S	6	100	88	11.6	5.1	6.5	56
DePaola et al. (1977)[18]	1000	1/day S	10–12	158	24	7.6	4.4	3.2	41

S = School Only

Table 6–2. Results of Clinical Trials of Mouthrinsing with Neutral Sodium Fluoride Solutions (1978–1982)

Study	F Concen- tration (ppm)	Rinsing Fre- quency	Age at Base- line	Number Com- pleting Trial	Length of Trial (Months)	DMFS Caries Increment Control	Test	Δ	% Differ- ence
Ripa et al. (1978)[19]	900	1/week S	7–12	750	24	3.2	2.6	*0.6*	20
Ringelberg et al. (1979)[20]	250	1/day S	11	179	30	6.3	4.8	*1.5*	23
Heifetz et al. (1982)[21]	900	1/week S	10–12	97	36	3.6	2.3	*1.3*	38
	900	1/week S	10–12	102	36	4.4	3.4	*1.0*	24
(Same trial)	225	1/day S	10–12	88	36	3.6	1.9	*1.7*	47
	225	1/day S	10–12	107	36	4.4	2.9	*1.5*	34
Driscoll et al. (1982)*[22]	900	1/week S	12–13	81	30	2.6	2.0	*1.6*	22
	900	1/week S	12–13	81	30	1.9	0.9	*1.0*	55
(Same trial)	225	1/day S	12–13	102	30	2.6	1.9	*0.5*	28
	225	1/day S	12–13	102	30	1.9	0.9	*1.0*	50

S = School only
*0.84 ppm F in drinking water

weekly mouthrinsing, however, the difference seems insufficient to warrant the increased logistical problems and cost involved in daily rinse programs. This was the conclusion from two studies that compared these regimens in the same trial,[21,22] and is, of course, an administrative rather than a biologic consideration.

Considering the number of different research teams and the variety of sources of study subjects, these consistent clinical results justify a high degree of confidence in the conclusion that rinsing with neutral sodium fluoride solutions is efficacious in the partial prevention of caries in children.

There are relatively few data available on the use of fluoride mouthrinsing by adults. One of the earliest reported trials of a sodium fluoride rinse used young adult subjects and failed to find evidence of cariostatic activity.[23] However, there is no known biological reason why caries-prone adults would not benefit from fluoride mouthrinsing. Rampant, postirradiation caries in adults has successfully been prevented by frequent application of fluoride gels,[24] and adults living in fluoridated communities have been observed to have a significantly lower prevalence of root-surface caries than those in a control town.[25] The National Institute of Dental Research is now conducting a clinical trial to test the efficacy of daily rinsing with sodium fluoride in persons aged 18 to 70 years.

In vitro enamel fluoride uptake is enhanced when the surface enamel is slightly demineralized by exposure to acid. This observation has led to the formulation of low pH fluoride rinses known as acidulated phosphate fluoride (APF). Table 6–3 summarizes the results of two clinical trials in which direct comparisons were made of the efficacy of daily APF and neutral pH rinses, which contained equal concentrations of the fluoride ion.[13,14] There was no significant advantage to the use of APF in these studies, and poor acceptance of the taste of the acidulated rinse was reported.[14] Other investigations[26,27] support the conclusion that APF rinses have about the same degree of cariostatic activity as neutral pH rinses.

Table 6–3. Results of Clinical Trials Comparing Neutral and Acidulated Phosphate Fluoride Mouthrinses

Study	F Concentration		Rinsing Frequency	Age at Baseline	Number Completing Trial	Length of Trial (Months)	DMFS Caries Increment			% Difference
							Control	Test	Δ	
Aasenden et al. (1972)[13]	200	NaF	1/day S	8–11	114	36	12.3	9.0	*3.3*	27
	200	APF	1/day S	8–11	109	36		8.7	*3.6*	
Heifetz et al. (1973)[14]	3000	NaF	1/day S	10–12	126	24	7.5	4.7	*2.8*	38
	3000	APF	1/day S	10–12	133	24		5.5	*2.0*	27

S = School only

Stannous Fluoride Mouthrinses

Stannous fluoride has been shown to be efficacious in caries prevention when used in dentifrices. There have been relatively few studies, however, of the preventive effects of this compound when used in a mouthrinse.

Table 6–4 shows the results of two clinical trials in which daily stannous fluoride rinses were tested at concentrations of 100, 200, and 250 ppm F.[28,29] In one of these studies, the subjects were also exposed to optimally fluoridated drinking water. Compared to placebo-treated control children, those rinsing with stannous fluoride developed about 20 to 40% fewer caries lesions during the trials, or about the same level of protection given by daily or weekly rinsing with sodium fluoride. Thus, the available clinical evidence indicates that the anticaries effect of stannous fluoride rinses is roughly the same as that of sodium fluoride rinses.

Other Fluoride Mouthrinses

Various other fluoride compounds have been tested as mouthrinses, but none has shown sufficient cariostatic activity, compared to sodium fluoride, to warrant their recommended use.

A clinical trial of an amine fluoride rinse showed no superiority over a neutral sodium fluoride rinse when used according to the same regimen.[30] In another investigation, an ammonium fluoride mouthrinse was no more effective than a sodium fluoride rinse when both were used daily in an acidulated form.

Table 6–4. Results of Clinical Trials of Mouthrinsing with Stannous Fluoride Solutions

Study	F Concentration (ppm)	Rinsing Frequency	Age	Number Completing Trial	Length (Months)	DMFS			% Difference
						Control	Test	Δ	
McConchie, et al. (1977)[28]	100	1/day S	10	199	36	5.8	4.6	*1.2*	20
	200	1/day S		204	36	5.8	4.8	*1.0*	17
Radike, et al. (1973)*[29] (Examiner #2)	250	1/day S	8–13	348	20	3.0	2.0	*1.0*	33
						2.8	1.6	*1.2*	43

S = School only
*Fluoridated water supply

Recently, claims have been made for the superiority of a combination of acidulated phosphate fluoride and stannous fluoride rinses, used either in sequence or as a mixture. These claims appear to be primarily based upon in vitro experiments, which have demonstrated that the sequential regimen is effective in reducing enamel solubility.[31] Enamel solubility reduction, however, has not been shown to be predictive of clinical efficacy. To date, there have been no clinical caries studies reported of the cariostatic effect of combinations of acidulated phosphate fluoride and stannous fluoride.

There is currently no basis for believing that this combination regimen will be therapeutically superior to the extensively tested sodium fluoride mouthrinses.

SAFETY OF FLUORIDE MOUTHRINSES

Mouthrinses pose no hazards when used according to recommended protocols. These include providing the correct amount of rinse solution, especially to young children, and adequate supervision to ensure that the rinse is not swallowed. Mouthrinsing is not recommended for preschool-aged children.

In a study of children aged 10 and 11 years, it was calculated that approximately 15% of a 10-ml mouthrinse was retained when the children followed a correct rinsing procedure.[32] With a rinse containing 0.2% NaF, this would result in the inadvertent ingestion of 1.35 mg F per week, far below any potentially toxic level. The corresponding calculation for a 0.05% NaF daily mouthrinse indicates an average inadvertent ingestion of less than 0.35 mg F/day.

Accidental or intentional swallowing of the entire volume of rinse will result in the ingestion of 9 mg F or 2.3 mg F for the weekly and daily rinse concentrations, respectively. Clearly, this is to be discouraged by proper instruction and supervision, even though this single potential dose of ingested fluoride is well below the minimum dose of 120 mg F estimated to be safely tolerated by a 5-year-old child.[33] Older children have a proportionally greater margin of safety.

Sensible precautions are necessary to prevent children from gaining access to large quantities of mouthrinse, or to the concentrated packages of fluoride often used to prepare rinses in school programs. Ingestion of an entire 2-gram packet of sodium fluoride, an amount commonly used in preparing rinses, would almost certainly have toxic results for all school children, and would most likely be lethal for young children.[33] As with all medicaments, large amounts of fluoride must be kept securely locked and inaccessible to children. With respect to fluoride mouthrinses sold for use at home, the largest available quantity dispensed contains about 118 mg F. Ingestion of the entire amount in a single dose is unlikely, as large amounts of sodium fluoride are emetic, however, ingestion would be expected to produce toxic symptoms in a young child.[33] Therefore, parents should be advised to exercise prudent supervision when fluoride mouthrinses are used at home.

Concern has been expressed regarding the safety of the combined use of acidulated phosphate fluoride and stannous fluoride as a mouthrinse. Following the recommendations of the manufacturer of one such product could result in the use of a rinse containing approximately 116 mg F, which if swallowed, would be likely to produce symptoms of acute fluoride intoxication in a young child.[34]

CONCLUDING REMARKS

Fluoride mouthrinses are a thoroughly tested and effective means for the partial prevention of caries in children. The available clinical evidence suggests that neutral sodium fluoride rinses are the formulation of choice, although the possible antiplaque activity of stannous fluoride rinses deserve further investigation. For children aged 6 years and above, rinsing should be done for 1 min using 10 ml of either a 0.2% NaF rinse once a week or a 0.05% NaF rinse daily.

Mouthrinsing is ideally suited for use in school-based preventive programs, as large numbers of children can be adequately supervised at minimal cost. All necessary supplies for school mouthrinse programs can be obtained for less than $1.00 per child per school year. The total cost of the program varied depending upon the supervision required, and ranged from $0.71 to $9.27 in 17 communities which participated in one demonstration program.[35]

The cost-effectiveness of mouthrinsing also depends upon the amount of caries actually prevented. This, in turn, is a function of the underlying caries susceptibility of the children involved. As with most preventive measures, fluoride mouthrinsing will appear less cost-effective the more successful the continued use of all preventive procedures are in reducing caries incidence among the children involved. This was evidently the case in one recently reported caries prevention demonstration program in which little benefit could be demonstrated for fluoride mouthrinsing compared to a reference group, as children in the latter group had very low increments of caries.[36]

Since 1974, increasing numbers of children have become involved in school-based fluoride mouthrinse programs in the United States. Recent estimates are that over 11 million children are participating each year in these programs, with an additional, unknown number now using fluoride rinses at home. The extent to which this preventive method has contributed to the recent decline in caries prevalence among children cannot be directly established. Since fluoride mouthrinsing has consistently been shown to provide about 35% protection against caries, we may assume that the widespread adoption of this method of caries prevention has had a substantial public health impact.

Fluoride mouthrinsing is a simple, well-accepted, safe, and relatively inexpensive adjunct to other methods of caries prevention, including adhesive sealants, fluoride dentifrices, and perhaps, community water fluoridation. Until it has been clearly demonstrated that caries incidence can be further decreased and maintained at low levels by other, simpler methods, fluoride mouthrinsing will continue to play a major role in efforts to prevent dental caries.

REFERENCES

1. Englander, H.R., et al.: Clinical anticaries effect of repeated topical sodium fluoride applications by mouthpieces. J. Am. Dent. Assoc., *100*:638–644, 1967.
2. Shern, R.J., et al.: Enamel biopsy results of children receiving fluoride tablets. J. Am. Dent. Assoc., *95*:310–314, 1977.
3. Jenkins, G.N., and Edgar, W.M.: Distribution and forms of F in saliva and plaque. Caries Res., *11*(Suppl. 1):226–242, 1977.
4. Hamilton, I.R.: Effects of fluoride on enzymatic regulation of bacterial carbohydrate metabolism. Caries Res., *11*(Suppl. 1):262–291, 1977.
5. Silverstone, L.M.: The effect of fluoride in the remineralization of enamel caries and caries-like lesions in vitro. J. Public Health Dent., *42*:42–53, 1982.
6. Larson, R.H., et al.: Caries inhibition in the rat by water-borne and enamel-borne fluoride. Caries Res., *10*:321–331, 1976.
7. Birkeland, J.M.: Direct potentiometric determination of fluoride in soft tissue deposits. Caries Res., *4*:243–248, 1970.
8. Yankell, S.L., et al.: Clinical effects of using stannous fluoride mouthrinses during a five-day study in the absence of oral hygiene. J. Periodont. Res., *17*:374–379, 1982.
9. Glantz, P.-O.: On wetability and adhesiveness. A study of enamel, dentin, some restorative dental materials, and dental plaque. Odont. Revy., *20*:1–132, 1969.
10. Horowitz, H.S. et al.: The effect on human dental caries of weekly oral rinsing with a sodium fluoride mouthwash: A final report. Arch. Oral Biol., *16*:609–616, 1971.

11. Brandt, R.S., et al.: The use of a sodium fluoride mouthwash in reducing the dental caries increment in eleven-year-old English school children. Proc. Br. Paedod. Soc., 2:23–25, 1972.
12. Moreira, B.H., and Tumang, A.J.: Council classifies fluoride mouthrinses. J. Am. Dent. Assoc., 91:1250–1251, 1975.
13. Aasenden, R., et al.: Effects of daily rinsing and ingestion of sodium fluoride solutions upon dental caries and enamel fluoride. Arch. Oral Biol., 17:1705–1714, 1972.
14. Heifetz, S.B., et al.: The effect on dental caries of weekly rinsing with a neutral sodium fluoride or an acidulated phosphate fluoride mouthwash. J. Am. Dent. Assoc., 87:364–368, 1973.
15. Rugg-Gunn, A.J., et al.: Caries prevention by daily fluoride mouthrinsing: report of a three-year clinical trial. Br. Dent. J., 135:353–360, 1973.
16. Gallagher, S.J., et al.: Self-application of fluoride by rinsing. J. Public Health Dent., 34:13–21, 1974.
17. Maiwald, H.J., and Padron, F.S.: Ergebnisse der kollectiven kariespravention durch Mundspulungen mit 0.2 prozentiger natrium fluoridlosung nach 88 Monaten. Stomatol. DDR, 27:835–840, 1977. (Eng. Abs.)
18. DePaola, P.F., et al.: Effect of high concentration ammonium and sodium fluoride rinses on dental caries in school children. Community Dent. Oral Epidemiol. 5:7–14, 1977.
19. Ripa, L.W., et al.: Supervised weekly rinsing with a 0.2 percent neutral NaF solution: Results from a demonstration program after two school years. J. Am. Dent. Assoc., 97:793–798, 1978.
20. Ringelberg, M.L., et al.: Caries-preventive effect of amine fluorides and inorganic fluorides in mouthrinses or dentifrice after 30 months of use. J. Am. Dent. Assoc., 98:202–208, 1979.
21. Heifetz, H.B., et al.: A comparison of the anticaries effectiveness of daily and weekly rinsing with sodium fluoride: Final results after three years. Pediatr. Dent., 4:300–303, 1982.
22. Driscoll, W.S., et al.: Caries-preventive effects of daily and weekly fluoride mouthrinsing in a fluoridated community: Final results after 30 months. J. Am. Dent. Assoc., 105:1010–1013, 1982.
23. Bibby, B.G., et al.: Preliminary reports on the effect on dental caries of the use of sodium fluoride in a prophylactic cleaning mixture and in a mouthwash. J. Dent. Res., 25:207–211, 1946.
24. Dreizen, S., et al.: Prevention of xerostomia-related dental caries in irradiated cancer patients. J. Dent. Res., 56:97–104, 1977.
25. Stamm, J.W., et al.: Comparison of root caries and/or fillings present in adults with lifelong residence in fluoridated and nonfluoridated communities. AADR Abstract #552, 1980.
26. Frankl, S.N., et al.: The topical anticariogenic effect of daily rinsing with an acidulated phosphate fluoride solution. J. Am. Dent. Assoc., 85:882–886, 1972.
27. Finn, S.B., et al.: The clinical cariostatic effectiveness of two concentrations of acidulated phosphate fluoride mouthwash. J. Am. Dent. Assoc., 90:398–402, 1975.
28. McConchie, J.M., et al.: Caries-preventive effects of two concentrations of stannous fluoride mouth rinse. Community Dent. Oral Epidemiol., 5:278–283, 1977.
29. Radike, A.W., et al.: Clinical evaluation of stannous fluoride as an anti-caries mouthrinse. J. Am. Dent. Assoc., 86:404–408, 1973.
30. Ringelberg, M.L., et al.: Effects of an amine fluoride dentifrice and mouthrinse on the dental caries in school children after 18 months. J. Prev. Dent., 5:26–30, 1978.
31. Shannon, I.L.: Antisolubility effects of acidulated phosphate fluoride and stannous fluoride in the treatment of crown and root surfaces. Aust. Dent. J., 16:240–242, 1971.
32. Birkeland, J.M.: Intra- and interindividual observations on fluoride ion activity and retained fluoride with sodium fluoride mouthrinses. Caries Res., 7:39–55, 1973.
33. Heifetz, S.B., and Horowitz, H.S.: The amount of fluoride in current fluoride therapies: Safety considerations for children. J. Dent. Child., (in press).
34. Horowitz, H.S., and Horowitz, A.M.: Letters to the editor. J. Public Health Dent., 41:6–7, 1981.
35. Miller, A.J., and Brunelle, J.A.: A summary of the NIDR community caries prevention demonstration program. J. Am. Dent. Assoc., 107:265–269, 1983.
36. Bell, R.M., et al.: Treatment effects in the national preventive dentistry demonstration program. Rand. Publ. No. R-3072-RWJ, Santa Monica, California, 1984.

Chapter 7
FLUORIDES IN PERIODONTAL THERAPY

Michael G. Newman, Dorothy A. Perry, Fermin A. Carranza Jr., and John E. Mazza

Caries and periodontal diseases are the most common infections in humans. The chronic nature and high prevalence of plaque-induced periodontal diseases create severe problems in relation to its treatment and prevention. Since plaque growth, with the potential to cause disease occurs constantly, long-term and continuous treatment is necessary. Using chemotherapy to prevent the growth of pathogens within plaque has had widespread acceptance and use. Among the most commonly used agents have been chlorhexidine and fluoride compounds.

Fluoride compounds have potential therapeutic, adjunctive, and preventive uses in the treatment of periodontal diseases. These compounds affect both the agents of disease and the tooth. Numerous studies over the past 15 years have suggested that stannous fluoride is among the best fluoride agents because it can reduce the following: (1) plaque and potentially, gingivitis; (2) bleeding upon probing in periodontal patients; (3) enamel solubility; (4) caries pathogens and potential periodontopathogens; and (5) dental hypersensitivity (sensitivity). Various formulations have been tested and the routes of their application have been investigated. Long-term studies are underway to determine the effect of fluoride agents over extended periods of time. Further research will clarify the role of stannous and other fluorides in periodontal disease treatment, control, and prevention.

Caries and periodontal diseases are the most common infections in humans. These infections result from the accumulation and growth of specific pathogenic bacteria within supra- and subgingival plaque.[1] In the prefluoride era, when caries rates were high, emphasis on dental treatment revolved primarily around the prevention and restoration of dental decay. Today, the nature, prevalence, and awareness of dental disease has changed dramatically. The National Institute of Dental Research estimates that 37% of American children have never had a cavity and the overall caries rate continues to decline.[2] The children of the fluoride era are now adults. Together with their preceeding generations, nine out of ten will be afflicted with periodontal diseases, and by age 60, one third of Americans will have lost all their teeth primarily because of gum disease.[3] These facts have created great interest in the diagnosis, treatment, and prevention of periodontal diseases within the dental community.

The chronic nature and high prevalence of periodontal diseases create severe problems relative to their treatment and prevention. Since plaque growth, with the potential to cause disease, occurs constantly, long-term and continuous treatment is necessary. It is now recognized that the targets of therapy are the specific agents of disease (specific pathogenic bacteria) and the tooth itself. Prevention of the growth

Table 7–1. General Properties of Topical Fluoride Agents in Periodontal Treatment*

Antiplaque	Tooth Effects
Supragingival	Remineralize
Subgingival	Strengthen
	Desensitize

Antibacterial	Potential Unwanted Effects
Selective	Increase resistant microorganisms
Non-selective	Ingestion–toxicity
Decrease recolonization of pathogens	Allergy

*Derived in part from references 4,6,10,13,16,19,49,54–58,59

of pathogens within the plaque by chemotherapeutic agents has had widespread interest and acceptance as an important and viable treatment modality. Among the most commonly used agents have been chlorhexidine and fluoride compounds. Recently discovered knowledge regarding the potential benefits of fluorides for use in treating periodontal diseases has set the stage for a new "fluoride era."

Fluoride compounds have potential therapeutic, adjunctive, and preventive uses in periodontal disease treatment. These compounds fulfill therapeutic goals by affecting pathogenic bacteria as well as teeth (Table 7–1). Numerous studies over the past 15 years have suggested that stannous fluoride (SnF_2) is among the best of the available fluoride agents because it can: (1) reduce plaque and potentially, gingivitis;[4–18] (2) reduce bleeding upon probing in periodontal patients;[19–21] (3) reduce enamel solubility;[22] (4) reduce caries pathogens[23–25] and potential periodontopathogens;[19–21,26,27] and (5) reduce dental hypersensitivity (sensitivity).[28] Various formulations have been tested and many routes of application have also been investigated (Table 7–2). Studies are underway to determine the effect of stannous fluoride and other fluoride agents over extended periods of time.

This chapter will briefly review: (1) the current concepts of periodontal disease, because the design, implementation, and use of fluoride agents depends upon an

Table 7–2. Modes of Delivery of Fluoride Agents in Periodontal Therapy

Delivery Modes	Agents**	Application	
		Patient*	Professional
Dentifrice	SnF_2, NaF, $Na_2 PO_3F$	×	—
Rinse	SnF_2, NaF, APF	×	×
Gel	SnF_2, NaF, APF	×	×
Varnish	Amine	—	×
Iontophoresis	SnF_2, NaF, APF	—	×
Floss	SnF_2, NaF	×	—
Waxes	SnF_2, NaF	—	×
Restorations	NaF	—	×
Irrigation	SnF_2, NaF, APF	×	×
Tooth Picks	SnF_2, NaF	×	—
Chewing Gum	SnF_2		

*Patient compliance is essential in optimizing the therapeutic effectiveness of these agents
**Not intended to be a comprehensive listing–SnF_2 = stannous fluoride; NaF = sodium fluoride; APF = acidulated phosphate fluoride; Amine = amine fluoride; $Na_2 PO_3F$ = sodium monofluorophosphate

understanding of the disease process; and (2) the current status of the use of fluorides, especially stannous fluoride, in periodontal disease treatment.

CURRENT CONCEPTS OF PERIODONTAL DISEASE

Progress in research in the microbiology and immunology of periodontal diseases has been dynamic and sophisticated.[29] This research has suggested that various types of periodontal diseases, as well as varying degrees of severity, are associated with different combinations of specific bacterial interactions within a complex host.

Supragingival Plaque. In gross anatomic terms, the formation of dental plaque occurs above (supragingival) and below (subgingival) the gingival margin (Fig. 7–1). Supragingivally, plaque formation initially involves the association of gram-positive bacteria with the tooth surface. Salivary and dietary components, oral hygiene, and local and host factors influence the nature and pathogenic potential of this plaque. Once supragingival plaque is established, successional acquisition of pathogenic bacteria as well as an increase in the amount of plaque are responsible for the ultimate development of gingivitis. Early bacteriologic changes in supragingival plaque initiate the inflammatory response in the gingiva. Most often, these changes are the result of poor or inadequate oral hygiene and include reactions such as edema, swelling, and an increase in gingival fluid. These physiologic and ana-

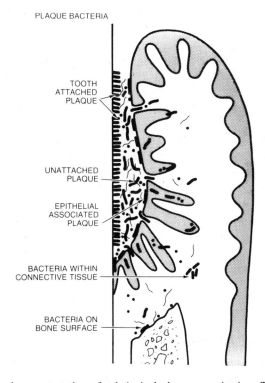

PLAQUE BACTERIA

TOOTH
ATTACHED
PLAQUE

UNATTACHED
PLAQUE

EPITHELIAL
ASSOCIATED
PLAQUE

BACTERIA WITHIN
CONNECTIVE TISSUE

BACTERIA ON
BONE SURFACE

Fig. 7–1. Diagrammatic representation of subgingival plaque organization. Tooth-associated or attached plaque contains primarily gram-positive organisms. The attached plaque is always covered with a layer of loosely adherent unattached plaque of predominantly gram-negative motile bacteria. This unorganized layer is attached to the pocket epithelial surface. Bacteria associated with the pocket surface may invade the epithelium and connective tissue and alveolar bone (not drawn to scale). Reprinted with permission from W.B. Saunders.[1]

tomic changes in the gingiva are partly responsible for the subsequent pathogenic alterations in the subgingival plaque. The successional changes in the bacteriology of supragingival (and subgingival) plaque suggest that specific bacteria are responsible for the observed "pathogenic" changes.[30,31] Reestablishment of a "young plaque" following treatment (whether mechanical alone or fluoride supplemented) is associated with a return to gingival health.[32]

Subgingival Plaque. Subgingival plaque is associated with destructive periodontal diseases as well as with disseminated infection. The bacteria that colonize this area are primarily anaerobic and capnophilic. Both gram-negative and gram-positive species are regularly isolated. Most of these bacteria utilize the protein and other nutrients provided in the subgingival environment by gingival fluid.

Specific subgingival bacteria and groups of bacteria have been implicated in several periodontal diseases, and some bacteria have been associated with periods of exacerbation and remission.[1] Subgingival anaerobes, capnophiles, and motile bacteria have been used by many authors as "markers" for determining the efficacy of periodontal therapy. Specifically, spirochetes and motile subgingival bacteria have been counted under the darkfield or phase contrast microscope, and other bacteria have been sought by the use of selective bacteriologic media.[46-47] Investigators have differentially quantitated the organisms isolated from scrapings of subgingival plaque before and after treatment (with or without chemotherapy). A reduction in the number of these bacteria has been suggested to coincide with improved clinical health.[46-48] How these groups of bacteria are actually involved in active disease is still questionable.

Recently, the subgingival plaque has been characterized on the basis of its anatomic association with the tooth surface[33,34] and adjacent soft and hard tissues[1,35-40] (Fig. 7–1). There are many important clinical and therapeutic implications concerning the bacterial colonization of the pocket. Periods of exacerbation in chronic and refractory forms of periodontitis may be initially caused by the invasion of subgingival pathogens.[45] The removal of these bacteria from within the pocket or the tissue by mechanical treatment, or the administration of antibacterial drugs has become the therapeutic goal of modern periodontal therapy.

Once established in the subgingival area, periodontal infection usually drains into the oral cavity via the periodontal pocket producing little, if any, discomfort or pain. Tooth mobility, bleeding gingiva, and increased spaces between the teeth are common but not necessarily signs of advanced disease. In many cases, deep-seated periodontal infection is marked by fibrotic gingiva or "normal" appearing gingiva.*

The recent microbiologic, anatomic, and immunologic advances regarding oral and odontogenic infections have provided an opportunity to begin testing therapeutic approaches aimed specifically at the suspected pathogens or the targets they destroy.[41-44]

FLUORIDES IN PERIODONTAL THERAPY

If fluorides are to be of value in periodontal therapy they must be shown to be safe, readily available, and effective against the agents of disease (microorganisms). Fluorides should have value in treating periodontally healthy patients (Type I), and

---

*It is imperative that a thorough clinical (periodontal) examination with a periodontal probe be given to all adult patients. Visual examination is inadequate.

patients with periodontal disease and those with arrested disease (Type II) (Table 7–3). The rationale for the clinical use of fluoride agents comes from extensive literature based on laboratory, animal, and human studies. For reviews, see Perry[49] and Tinanof.[16] Recent information on the potential role of fluoride agents in the treatment of Type I and Type II patients follows.

Type I–Periodontal Health. Type I patients are generally periodontally healthy (Table 7–3). The treatment objective is simple—*prevent disease*. Besides the anti-caries effects and the possible reduction in dental sensitivity (hypersensitivity), the major role of fluorides is to *prevent* the establishment of pathogens within the plaque. This can be accomplished in several ways and includes: (1) a reduction in plaque volume and age; the older and more copious the plaque, the more likelihood that pathogens will colonize; (2) a reduction in the deleterious effects of pathogens in existing plaques, thereby maintaining periodontal health; (3) the retention of fluoride within the plaque and/or the tooth for potentially prolonged beneficial effects (substantivity).

Type I periodontal patients do not require special or unusual fluoride therapy, such as the use of long-term high-concentration agents. As long as these patients receive the benefits of fluoridated toothpastes, rinses, or gels, prevention of disease and maintenance of health are accomplished by regular visits to the dental health care professional for mechanical plaque and calculus removal and oral hygiene reinforcement. It is suggested that fluoride agents (which must be shown to be safe over extended periods of time) may improve the efficiency of oral hygiene, decreasing the likelihood of developing periodontitis.

Type II–Active or Maintenance Periodontal Therapy. The Type II patient is either undergoing active or maintenance periodontal therapy. The rationale for the current use of fluorides in periodontal treatment is based on the knowledge that all common forms of periodontal disease are caused by plaque bacteria. *While undergoing active treatment, the goal of fluoride therapy is to augment mechanical treatment modalities.* Augmentation may be accomplished by: (1) killing pathogenic bacteria within the plaque; (2) preventing or slowing the recolonization of periodontopathogens within the plaque (and perhaps subsequently within the tissues); (3) reducing plaque volume; (4) decreasing post-treatment sensitivity; (5) decreasing recurrent, root, or new caries.

Table 7–3. Types of Periodontal Patients

(Type I) Periodontally Healthy Patient
No history of destructive disease (bone loss)
Minimum local gingivitis
Minimum local dental sensitivity
Regular dental check-ups (maintenance)

(Type II) Periodontal Disease Patient
Active disease present
Signs of disease: pockets, mobility, bone loss, recession, root caries
Arrested (treated) disease (periodontal maintenance)

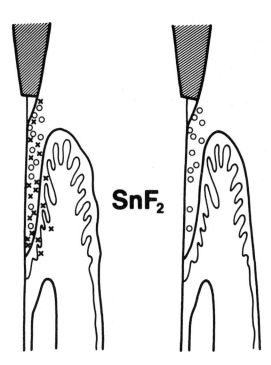

Fig. 7–2. (Left) Periodontal pocket with associated supra- and subgingival plaque. Oral surface of gingiva depicts clinical inflammation. 0 = health-associated bacteria, × = periodontopathogens. (Right) Use of stannous fluoride (SnF_2) results in killing periodontopathogens (×) and recolonization by health-associated bacteria (0). Resolution of inflammation and return to health results from adjunctive antimicrobial effects of SnF_2. The agent may also reduce caries and hypersensitivity of the root surface.

Antibacterial Effects Within Plaque. Topical antibacterial agents have been used experimentally and clinically in periodontal therapy for many years. In addition to fluoride-containing agents, tetracycline, chlorhexidine, alexidine, hydrogen peroxide, sodium bicarbonate, and a variety of other agents have been used. None of the agents permanently eliminated plaque. Therapeutic effectiveness comes from the ability to kill pathogens that are present and to prevent or slow recolonization. The targets of these agents have been both supra- and subgingival plaque bacteria.[26,27,50–53]

Supragingival plaque alteration augmented by antimicrobial agents can retard the growth of pathogenic bacterial recolonization not only in the supragingival plaque but in the subgingival plaque as well. This may occur in at least three ways: (1) the agents may directly kill the bacteria; (2) the agents may reduce plaque volume; and (3) the agents may prevent the downward growth of supragingival plaque into the subgingival environment, where it can grow farther away from the effects of topical agents (Fig. 7–2). Since the successional changes responsible for the establishment of pathogens usually occur in mature plaque, keeping the plaque "young" will keep the plaque compatible with health. This is also the effect of oral hygiene procedures not employing adjunctive chemotherapeutic agents. Thus, the major issue is whether antimicrobial agents have a significant effect on the plaque beyond those of mechanical cleaning.

The antibacterial potential for fluorides against pathogens has been demonstrated.

Yoon et al.[26] examined fluoride effects on "*Actinomyces*-like" organisms from 11 adults, since these organisms are related to root caries and gingivitis. It was reported that sodium fluoride, stannous fluoride, and acidulated phosphate fluoride (APF) were all effective in decreasing the percentage of this group of bacteria. In vitro antimicrobial susceptibility studies demonstrated that stannous fluoride had the most effect in one hour, sodium fluoride the least; at 24 hours, stannous fluoride had the most inhibitory effect, and APF and sodium fluoride were less but similarly effective.

Yoon et al.[27] determined the effects of sodium fluoride, stannous fluoride, and APF in vitro on the subgingival gram-negative organisms *Bacteroides melaninogenicus*, subspecies *melaninogenicus* and *asaccharolyticus*. Stannous fluoride was shown to be more effective at lower concentrations and at shorter times than APF or sodium fluoride. Stannous fluoride was more acidic than APF, and its effectiveness was increased by this property, but acidity was shown not to be exclusively responsible for bacterial inhibition in vitro. From this and numerous other studies,[16,49] it has been determined that the stannous ion plays a major role as an antibacterial agent. Stannous ions adsorb to the surfaces of microorganisms where they can influence metabolism and subsequently pathogenicity. In addition, alterations in the surface charge of gram-positive cells can have an effect on the surface binding capability, thereby decreasing the potential for retention within the plaque. The *exact* mechanisms of action are not completely understood.

The preceding in vitro studies suggested that stannous fluoride agents aimed at potential pathogenic bacteria could effectively reduce the numbers of bacteria and thereby have the potential to control their associated diseases. Clinical confirmation of these effects has been demonstated, but data are still scant.

Mazza et al.[19] demonstrated that 1.64% SnF_2 was more effective that 0.4% SnF_2 or saline in reducing motile bacteria and spirochetes from subgingival plaque samples of ten adult males with advanced periodontitis. Solutions were irrigated into subgingival sites with a calibrated tuberculin syringe to 1 mm less than probing depth. Bacterial counts were taken over ten-week periods, and a solution of 1.64% SnF_2 reduced the count by 48.6% the first week and maintained a 7.5% reduction at week ten. This preparation also caused the most dramatic and long-lasting reduction of bleeding at the experimental sites; 0.4% SnF_2 showed a lesser effect, while the saline solution had no effect on bleeding.

Perry et al.[20] sought to determine the effect of subgingivally applied stannous fluoride as an adjunct to scaling and root planing in Type II periodontal patients. Following full mouth scaling and root planing, 1.64% SnF_2 was irrigated into periodontal pockets of 6-mm depth in one quadrant. (Similarly involved control sites in other quadrants received either saline lavage following treatment or no agent.) Patients were monitored clinically and microbiologically at 1, 3, 7, 12, and 16-week periods following treatment. The results demonstrated that in 6-mm treated pockets, 1.64% SnF_2 significantly reduced post-treatment colonization of potential pathogens up to 7 weeks. Mazza et al.[21] investigated the role of 0.4% SnF_2 in Type II patients who were *not* maintaining periodontal health at regularly scheduled 3-month maintenance (recall) visits in a private practice setting. Patients who demonstrated generalized bleeding upon probing in three out of four quadrants received traditional mechanical therapy plus a 3-month supply of supplementary 0.4% SnF_2 brush-on gel to be used on a daily basis in place of toothpaste in their oral hygiene routines. Patients in the experimental (SnF_2) group had significant reductions in supragingival plaque and gingival bleeding after 3 and 6 months of use. No deleterious effects were noted in either study.

Wieder et al.[14] combined the use of self-administered subgingivally applied chlorhexidine combined with a stannous fluoride dentifrice. Their findings suggested that "control" of both supra- and subgingival plaque bacteria by this chemotherapeutic approach had a dramatic and sustained effect of decreasing the signs of periodontal disease.

Studies such as these clearly demonstrate the potentially powerful adjunctive effects of topical fluorides (particularly SnF_2) in periodontal therapy. Since these agents are adjunctive, they must be combined with routine therapeutic measures, which include proper instruction in oral hygiene, adequate mechanical treatment for the removal of supra- and subgingival plaque and calculus, and periodic mandatory monitoring of the patient's dental health status. Without *all* of these measures, the prevention of disease will not predictably occur.

Further work is needed to determine the role of stannous fluoride and other fluorides in the treatment and prevention of periodontal diseases. Some of the many issues remaining to be explored are: (1) the most effective treatment regimens for adjunctive agents; (2) the type of monitoring required to measure effectiveness; (3) the long-term effects; (4) the identification of optimum doses and agents for each indivdual periodontal patient's needs; and (5) the demonstration of possible treatment alternatives not requiring patient compliance.

CONCLUDING REMARKS

The chronic nature and high prevalence of periodontal diseases have created increased awareness and an increased need for improved periodontal therapy. Since all forms of periodontal disease are caused by plaque bacteria, the use of adjunctive antimicrobial agents is a promising therapy. Fluorides have had a long, safe, and successful role in dentistry. In the past, they have primarily been used to reduce caries incidence. Currently, stannous fluoride has been shown to be an effective adjunct to routine periodontal procedures because it can reduce the amount of plaque, has demonstrated bacteriocidal and bacteriostatic action, and has reduced some signs of clinical disease.

REFERENCES

1. Newman, M.G., and Saglie, R.S.: The role of microorganisms in periodontal disease. *In* Clinical Periodontology. 6th Ed., Chapter 25. Edited by F.A. Carranza. Philadelphia, W.B. Saunders Co., 1984.
2. ADA News: Special Supplement. March 12, 1984.
3. Coady, J.M.: "ADA letter to colleagues" for Periodontal Disease: Don't Wait Til it Hurts. ADA News, March 12, 1984.
4. Yankell, S.L., et al.: Clinical effects of using stannous fluoride mouthrinses during a five-day study in the absence of oral hygiene. J. Periodont. Res., *17*:374–379, 1982.
5. Yankell, S.L., et al.: Effects of topically applied stannous fluoride and acidulated phosphate fluoride alone and in combination on dental plaque. J. Periodont. Res., *17*:380–383, 1982.
6. Leverett, D.H., et al.: The effect of daily mouthrinsing with stannous fluoride on dental plaque formation and gingivitis–four month results. J. Dent. Res., *60*:781–784, 1981.
7. Hock, J., and Tinanoff, N.: Resolution of gingivitis in dogs following topical application of 0.4% stannous fluoride and toothbrusing. J. Dent. Res., *58*:1652–1653, 1979.
8. Ellingsen, J.E., et al.: The effects of stannous and stannic ions on the formation and acidogenicity of dental plaque in vivo. Acta Odontol. Scand., *38*:219–222, 1980.
9. White, S.T., and Taylor, P.P.: The effect of stannous fluoride on plaque scores. J. Dent. Res., *58*:1850–1852, 1979.

10. Bay, I., and Rolla, G.: Plaque inhibition and improved gingival condition by use of a stannous fluoride toothpaste. Scand. J. Dent. Res., *88*:313–315, 1980.
11. Gross, A., and Tinanoff, N.: Effect of SnF_2 mouthrinse on initial bacterial colonization of tooth enamel. J. Dent. Res., *56*:1179–1183, 1977.
12. Tinanoff, N., et al.: Effect of stannous fluoride mouthrinse on dental plaque formation. J. Clin. Periodontol., *7*:232–241, 1980.
13. Svantun, B., et al.: A comparison of the plaque-inhibiting effect of stannous fluoride and chlorhexidine. Acta Odontol. Scand., *35*:247–250, 1977.
14. Wieder, S.G., et al.: Stannous fluoride and subgingival chlorhexidine irrigation in the control of plaque and chronic periodontitis. J. Clin. Periodontol., *10*:172–181, 1983.
15. Yankell, S.L., et al.: Effects of topically applied stannous fluoride and acidulated phosphate fluoride alone and in combination on dental plaque. J. Periodont. Res., *17*:374–379, 1982.
16. Tinanoff, N., and Weeks, D.B.: Current status of SnF_2 as an antiplaque agent. Pediatr. Dent., *1*:199–204, 1979.
17. Hochenedel, A.M., et al.: Prevention of plaque formation in preschool children by daily brushing with 0.4% stannous fluoride gel: A feasibility study. Tex. Dent. J., *26*:6–9, 1982.
18. Hellden, L., et al.: Clinical study to compare the effect of stannous fluoride and chlorhexidine mouthrinses on plaque formation. J. Clin. Periodontol., *8*:12–16, 1981.
19. Mazza, J.E., et al.: Clinical and antimicrobial effect of stannous fluoride on periodontitis. J. Clin. Periodontol., *8*:203–212, 1981.
20. Perry, D.A., et al.: Stannous fluoride adjunct to root planing, clinical, and antimicrobial effects. J. Dent. Res. 63:(Abstract 702), 1984.
21. Mazza, J., et al.: The effect of daily self-applied SnF_2 on clinical parameters of periodontitis. J. Dent. Res., *63*:(Abstract 876), 1984.
22. Shannon, I.L.: In vitro enamel solubility reduction through sequential application of acidulated phosphofluoride and stannous fluoride. J. Can. Dent. Assn., *36*:308–310, 1970.
23. Zickert, I., and Emilson, C.G.: Effect of a fluoride-containing varnish on *Streptococcus mutans* in plaque and saliva. Scand. J. Dent. Res., *90*:423–428, 1982.
24. Svanberg, M., and Westergren, G.: Effect of SnF_2, administered as mouthrinses or topically applied, on *Streptococcus mutans, Streptococcus sanguis,* and lactobacilli in dental plaque and saliva. Scand. J. Dent. Res., *91*:123–129, 1983.
25. Keene, H.J., et al.: Effect of multiple dental floss SnF_2 treatment on *Streptococcus mutans* in interproximal plaque. J. Dent. Res., *56*:21–27, 1977.
26. Yoon, N.A., et al.: The antimicrobial effect of fluorides (acidulated phosphate, sodium, and stannous) on *Actinomyces viscosus*. J. Dent. Res., *58*:1824–1829, 1979.
27. Yoon, N.A., et al.: Antimicrobial effect of fluorides on *Bacteroides melaninogenicus* subspecies and *Bacteroides asaccharolyticus*. J. Clin. Periodontol., *7*:489–494, 1980.
28. Thrash, W.J., et al.: A method to measure pain associated with hypersensitive dentin. J. Periodontol., *54*:160–162, 1983.
29. Löe, H.: Closing remarks: Microbiological and immunological aspects of oral diseases. J. Dent. Res., *63*:476–477, 1984.
30. Holdeman, L.V., Burmeister, J.A., and Moore, W.E.C.: Bacteriology of human experimental gingivitis. II. Species of interest. J. Dent. Res., *63*:349(Abstract 1538), 1982.
31. Moore, W.E.C., Good, I.J., and Hash, D.E.: Bacteriology of human experimental gingivitis. I. Statistics and bacterial ecology. J. Dent. Res., *61*:349 (Abstract 1537), 1982.
32. Mousques, T., Listgarten, M.A., and Phillips, R.W.: Effect of scaling and root planing on the composition of the human subgingival microbial flora. J. Periodont. Res., *15*:144–151, 1980.
33. Listgarten, M.A.: Structure of the microbial flora associated with periodontal health and disease in man: A light and electron microscopic study. J. Periodontol., *47*:1–18, 1976.
34. Newman, H.N.: The apical border of plaque in chronic inflammatory periodontal disease. Br. Dent. J., *141*:105–113, 1976.
35. Saglie, R., et al.: Bacterial invasion of gingiva in advanced periodontitis in humans. J. Periodontol., *53*:217–222, 1982.
36. Allenspach-Petrzilka, G.E., and Guggenheim, B.: *Bacteroides melaninogenicus* subspecies *intermedius* invades rat gingival tissue. J. Dent. Res., *61*:259 (Abstract 728), 1982.
37. Fillery, E.D., and Pekovic, D.D.: Identification of microorganisms in human gingivitis. J. Dent. Res., *61*:253 (Abstract 675), 1982.
38. Gillett, R., and Johnson, N.W.: Bacterial invasion of the periodontium in a case of juvenile periodontitis. J. Clin. Periodontol., *9*:93–100, 1982.
39. Frank, R.M.: Bacterial penetration in the apical pocket wall of advanced human periodontitis. J. Periodont. Res., *15*:563–573, 1980.
40. Saglie, R., et al.: Scanning electron microscopy of the gingival wall of deep periodontal pockets in humans. J. Periodont. Res., *17*:284–293, 1982.

41. Loe, H., and Korman, K.: Strategies in the use of antibacterial agents in periodontal disease. *In* Host-parasite Interactions in Periodontal Disease. Edited by R.J. Genco, and S. Mergenhagen. Washington, D.C., American Society for Microbiology, 1982, pp. 376–381.

42. Socransky, S.S.: Criteria for the infectious agents in dental caries and periodontal disease. J. Clin. Periodontol., *6*(Supplement):16–21, 1979.

43. Lindhe, J.: Treatment of localized juvenile periodontitis, clinical implications. *In* Host-parasite Interactions in Periodontal Disease. Edited by R.J. Genco, and S. Mergenhagen. Washington, D.C., American Society for Microbiology, 1982, pp. 382–394.

44. Slots, J., et al.: Periodontal therapy in humans. I. Microbiological and clinical effects of a single course of periodontal scaling and root planing, and of adjunctive tetracycline therapy. J. Periodontol., *50*:495–509, 1979.

45. Robertson, M., et al.: Correlation of bacterial invasion and disease activity. J. Dent. Res., *63*: (Abstract 470), 1984.

46. Listgarten, M.A., and Hellden, L.: Relative distribution of bacteria at clinically healthy and periodontally diseased sites in humans. J. Clin. Periodontol., *5*:115–132, 1978.

47. Listgarten, M.A.: Colonization of subgingival areas by motile rods and spirochetes: Clinical implications. *In* Host-parasite Interactions in Periodontal Disease. Edited by R.J. Genco, and S. Mergenhagen. Washington, D.C., American Society for Microbiology, 1982, pp. 112–120.

48. Rosling, B.G., and Slots, J.: Topical chemical antimicrobial therapy in the management of the subgingival microflora and periodontal disease. J. Dent. Res., *61*:273 (Abstract 854), 1982.

49. Perry, D.A.: Fluorides and Periodontal Disease: A Review of the literature. J. West. Soc. Perio., *30*:93–105, 1982.

50. Andres, C.J., et al.: Comparison of antibacterial properties of stannous fluoride and sodium fluoride mouthwashes. J. Dent. Res., *53*:457–460, 1974.

51. Mandell, R.L.: Sodium fluoride susceptibilities of suspected periodontopathic bacteria. J. Dent. Res., *62*:706–708, 1983.

52. Brown, L.R., et al.: Effect of continuous fluoride gel use on plaque fluoride retention and microbial activity. J. Dent. Res., *62*:746–751, 1983.

53. Trieger, N., and Chomenko, A.: New Concepts in the treatment of periodontitis. Oral Maxillofac. Surg., *43*:701–708, 1982.

54. Gabler, W.L., et al.: Fluoride inhibition of polymorphonuclear leukocytes. J. Dent. Res., *58*:1933–1939, 1979.

55. Hock, J., et al.: Blood and urine fluoride concentrations associated with topical fluoride applications on dog gingiva. J. Dent. Res., *60*:1427–1431, 1981.

56. Gabler, W.L., and Leong, P.A.: Effect of fluoride on polymorphonuclear leukocyte myeloperoxidase. J. Dent. Res., *59*:135, 1980.

57. Holland, R.I.: Cytotoxicity of fluoride. Acta Odontol. Scand., *38*:69–79, 1980.

58. Solheim, H., et al.: Chemical plaque control and extrinsic discoloration of teeth. Acta Odontol. Scand., *38*:303–309, 1980.

59. Stoller, N.H., et al.: Clinical evaluations of an amine fluoride mouthrinse on gingival inflammation and plaque accumulation. J. Periodontol., *48*:650–653, 1977.

Chapter 8
IMPACT OF FLUORIDES ON ROOT CARIES AND ROOT DENTINAL SENSITIVITY

Ernest Newbrun

The studies that have been described involving topical and systemic fluoride (F) in humans and rats consistently point to an inhibiting effect on root caries. However, much of the information is limited in scope and has only been reported in preliminary form. The epidemiological data compare only two levels of fluoride in the water supply and do not include a population that drinks an optimal concentration. The topical fluoride studies on root caries prevention have of necessity been limited in numbers of participants, extent of compliance, and duration of observations. Carefully designed studies are still needed to establish which mode of fluoride therapy will be most effective and will gain satisfactory compliance.

Sensitivity of root dentin to thermal, osmotic, chemical or tactile stimuli occurs when the dentinal tubules are exposed. According to the hydrodynamic theory of dentinal sensitivity, the fluid in these tubules transmits pressure changes to mechanoreceptors in the pulp causing the sensation of pain. The goal of treatment for desensitization therefore has usually been to block the tubules mechanically. Such blockage may occur naturally by components of the saliva or gingival crevice fluid or by ingredients in self-applied dentifrices. Many patients with sensitive dentin improve spontaneously without therapeutic intervention. For those patients who suffer from severe discomfort, which can interfere with eating, drinking, and oral hygiene, effective dentinal desensitization providing long-lasting relief is essential. Fluoride compounds, including sodium fluoride (NaF), stannous fluoride (SnF_2), and sodium monofluorophosphate ($Na_2 PO_3F$), have been used both professionally and self-applied for desensitization and are reported to be clinically effective. Other agents besides fluorides have also been found to decrease dentinal sensitivity. A critical review of many of these studies reveals that there were no controls, no double blind protocol, and only the subjective reports of the participants on which to base improvement. A large placebo effect is evident in strictly subjective evaluations.

Current demographic data showing that the life expectancy of Americans is increasing mean that the proportion of elderly people in the population is steadily growing larger. At the same time, dental surveys in the United States and other Western industrialized countries have shown a decline in the prevalence of coronal caries. These two trends, an increase in longevity and a decrease in loss of teeth due to caries, suggest that the treatment and management of periodontal diseases will become an increasingly important component of dental care. One consequence of periodontal diseases is the apical migration of the gingival attachment. This event, the denudation of the root surface, is a prerequisite of both root caries and root dentinal

93

sensitivity. For both of these conditions, precise information concerning epidemiology and etiology is scarce. Fluorides have been used, often empirically, for the prevention of root caries and the treatment of root dentinal sensitivity.

EFFECT OF TOPICAL AND SYSTEMIC FLUORIDE ON ROOT CARIES

Caries-preventive treatment in patients at high risk for root caries may involve oral hygiene instruction and dietary advice, but the cornerstone of any preventive regimen for these patients should include some mode of fluoride therapy. One such high-risk group are those whose salivary flow has been reduced or eliminated altogether, whether due to pathological changes in the salivary glands or exposure to therapeutic irradiation. Two forms of fluoride treatment have been advocated and tested for this population. One form of treatment uses an acid or neutral gel containing 1.1% NaF, which provides 5000 ppm F and is applied to the mouth in a custom-fitted soft plastic tray for 5 to 10 min/day.[1-3] The other form uses 0.4% SnF_2 gel, which provides 1000 ppm F,[4] and is usually applied directly on the teeth with a toothbrush for about 1 min/day for patients with mild to moderate xerostomia. This fluoride treatment can also be applied in a custom-fitted tray for 5 to 10 min/day for patients with severe xerostomia.[5]

The effectiveness of a 1.1% NaF gel applied in a custom-fitted polyvinyl mouth piece was originally demonstrated with children in a supervised school program,[6] but has not been widely adopted for children because it is not cost-effective. However, for a high risk adult population who are no longer in the mixed dentition stage, it may well be cost-effective. In one study a 1.1% NaF gel was used daily for up to 6.5 years by 67 irradiated patients, and only 30% developed caries.[1] The same gel applied daily had a profound effect in inhibiting the development of radiation caries in 42 xerostomic patients aged 17 to 76.[2] The caries increment of this treatment group was 0.07 DMFT/month (or 0.84 DMFT/year), whereas patients in a control population who were only using oral hygiene developed such rapid caries that after 3 months they were also placed on the 1.1% NaF regimen. Patients not using the 1.1% NaF gel developed immediate and pronounced increases of *Streptococcus mutans* in their dental plaque. Daily self-application of this gel for 5 min, while not eliminating this cariogenic flora, significantly reduced the increase of *S. mutans* that followed irradiation. Furthermore, this mode of fluoride therapy significantly reduced acid production in plaque.[7]

The preceding studies used a self-applied 1.1% NaF gel providing 5000 ppm F. An entirely different regimen was used by Johansen and Olsen[3] on 155 patients with high caries susceptibility due to a variety of problems, including therapeutic irradiation, diminished saliva, and defective tooth structure. Patients were provided with custom-made trays of soft plastic and either an acidulated phosphate fluoride (APF) gel containing 1.23% F (12,300 ppm) or, for those whose roots or dentin were too sensitive, a neutral sodium fluoride gel containing 1% F (10,000 ppm). These high-concentration fluoride gels are normally used only for professional application in the dental office and not for home use. The treatment schedule consisted of two 5-min applications each day for 2 weeks, followed by a once-daily application for an additional 2 weeks. After each application, patients rinsed for 2 min with a "remineralizing mouthwash" containing 5 mM calcium, 3 mM phosphate, and 0.25 mM fluoride (5 ppm). When the 4 weeks of fluoride gel treatment were over, the mouthwash was continued for the duration of the 3-year study period. Caries increments of 0.2, 0.2, and 0.3 new lesions/patient/year were reported for the first, second, and third year respectively. These investigators claimed that self-applied fluoride treatments could be discontinued after just 4 weeks if followed by a regimen consisting of proper oral hygiene, fluoride toothpaste, daily use of a supersaturated remineralizing mouthwash, and salivary stimulation by chewing gum.

The 0.4% SnF_2 gel has been adopted for daily self-treatment in the Veterans Administration Hospitals. Wescott et al.[4] reported using this regimen on 24 patients who had received irradiation for malignant lesions of the head and neck. In 6 patients who used the gel daily, no crowns were amputated and only one carious area was found during 3.75 years. Nine patients who either refused to use the gel or used it sporadically had 57 crowns amputated and an additional 75 carious surfaces during 3.75 years. Nine patients were lost to follow-up. The high proportion of noncompliance can be ascribed in part to the unpleasant taste of the original 0.4% SnF_2 gel formulated by the VA Hospitals. Commercially available 0.4% SnF_2 gels have been formulated with more acceptable flavors that would encourage better compliance. Five 0.4% SnF_2 products have been accepted by the Council on Dental Therapeutics of the ADA (Flo-Gel, Gel-Kam, Omnii, Gel Tin, STOP), and one additional product (Easy Gel) is presently undergoing review by the Council. The caries prevention program used at the University of Texas M.D. Anderson Hospital since 1980 involves the daily topical application of a 0.4% SnF_2 gel.[8] Stannous fluoride gel has been found to reduce *Streptococcus mutans* levels in the plaques of xerostomic patients better than sodium fluoride, but neither agent prevented the increase of lactobacilli in the postradiotherapy period.[5]

Self-applied sodium fluoride mouthrinses (either 0.2% used for 1 min weekly or biweekly, or 0.05% used for 1 min daily) have been amply documented as effective in reducing coronal caries in children.[9-11] Only one study has been reported on the use of an acidulated 0.1% NaF mouthrinse (pH 4, 452 ppm F) in adults (dental students) who rinsed 3 times/week for 1 year.[12] The duration of rinsing was not stated, but presumably was uncontrolled and brief. No benefit could be demonstrated, but this finding is inconclusive as the number of participants was too small (subjects who did not comply with the regimen because of the rinse's unpleasant taste were eliminated from consideration). A self-applied APF rinse (Phos-Flur Oral Rinse Supplement) containing 0.044% NaF has been routinely used to treat postirradiation xerostomic patients in the Oral Medicine Clinic at the University of California, San Francisco. Patients who observe this regimen develop fewer caries than those who are noncompliant,[13] but no data are available to document the benefits of such a rinse program.

The dose-response relationship for topical fluorides is unclear. In a fluoride dentifrice study on children, there was a dose-response caries-preventive effect with sodium fluoride dentifrices containing 250, 750, and 1000 ppm of F.[14] More recently, Koch et al.[15] claimed no difference in caries prevention between children who used a dentifrice with 250 ppm F and those using dentifrices with 1000 ppm F. It is, therefore, not possible to predict whether the 1.1% NaF gel (5000 ppm F) would be more effective than a 0.4% SnF_2 gel (1000 ppm F) or a 0.05% NaF rinse (226 ppm F) without direct clinical testing.

In a comprehensive review of the literature on the use of topical fluorides to prevent dental caries in adults, Swango[16] noted that few of the studies cited above have been replicated. It is therefore difficult to discern a trend of success for any given regimen. There has been no direct comparison of the efficacy of a 1.1% NaF gel with a 0.4% SnF_2 gel, although these two vehicles are the most frequently advocated. Swango[16] also observed that there are no adequate existing studies on the effect of topical fluoride procedures in inhibiting the development of root caries.

The studies reviewed thus far concern the use of topical fluoride to prevent root caries. Topical sodium fluoride, either self- or professionally applied, has also been tested, together with mechanical smoothing of the root surface, in efforts to arrest incipient or shallow root caries lesions. At the University of Texas in Houston, investigators classified root caries as follows: Grade I, incipient; Grade II, shallow surface defect, some pigmentation; Grade III, deep lesion; and Grade IV, pulpal involvement. Grade I lesions were treated by topical fluoride gel alone, using 1.0% NaF daily in custom-fitted trays. Grade II lesions were either recontoured, smoothed, and treated with fluoride, or treated only with fluoride. After 6 months, 15/20 Grade I lesions showed no clinical changes (visual or tactile), 3/20 progressed to Grade II, and 2/20 were arrested. All Grade II lesions (13/13) that were treated mechanically and with fluoride were clinically sound. The Grade II lesions (5/15) treated with fluoride alone showed no change in status.[17] In a similar study at the University of Alabama, root caries lesions were excavated, recontoured, and treated with 0.5% I_2, 1% KI and 1.2% NaF at 0, 9, and 16 days. In addition, patients used a 0.2% NaF mouthrinse daily.[18] In both these studies, levels of *Streptococcus mutans* on the root surface lesions were significantly reduced below pretreatment levels. However, as little is known about the specific role of *S. mutans* in root caries, the importance of this reduction remains to be established.

Just as the data on topical fluoride therapy in adults is limited, so is information on the effect of systemic ingestion of fluoride on root caries in adults. The only epidemiological data available indicate that lifelong residence in a fluoridated community is associated with a highly significant reduction of the prevalence of root caries or root fillings at all ages. Stamm and Banting[19] compared root caries lesions and fillings in 465 adults (mean age 42.8) residing in Woodstock, Ont., where the water contains 0.1 ppm F, with those of 502 adults (mean age 40) in Stratford, Ont., which has 1.6 ppm F in the water. The mean number of decayed or filled root surfaces was 1.36 and 0.64, respectively.

Root caries in rats is also significantly reduced by systemic administration of fluoride in the drinking water. Rats given 0, 4.5 or 45 ppm F developed root surface caries scores of 19.5, 11.8, and 0.35, respectively.[20] More recently, Rosen et al.[21] have found that topical fluoride swabbed twice a day on the molar teeth of rats also significantly inhibited root surface caries in comparison with a water control. The agents tested were a sodium fluoride solution (5000 ppm F), a sodium fluoride dentifrice slurry (500 ppm F), and water (0 ppm F). The corresponding root caries scores were 10.4, 15.5, and 24.1.

DENTINAL SENSITIVITY

When the root surface is denuded by the loss of cementum and periodontal tissues, dentinal tubules become exposed. Such exposure results in pain for many patients after thermal, osmotic, chemical or tactile stimulation. The term "hypersensitive" dentin is probably a misnomer, as it denotes excessive or above normal sensitivity. In fact, it is normal for almost all dentin to be sensitive when newly exposed, particularly if there has been rapid exposure, such as that following fracture of a tooth, scaling and root planing, or periodontal surgery. Slower exposure of the dentin, which occurs from abrasion, erosion, or simply gingival recession, is less likely to result in sensitive dentin. Similarly, many patients whose dentin is initially sensitive will experience decreased sensitivity or even "spontaneous" loss of sensitivity with time.

One can better evaluate the effectiveness of the various therapeutic agents that have been used for desensitization if one understands how pain is transmitted in dentin. Currently the consensus is that somatic nerve fibers are found only in the deepest parts of the coronal dentin-pulp interface under cusps and are confined to predentin in roots,[22,23] and that the odontoblastic processes generally are confined to the inner pulpal 25% of the tubule's length (see Fig. 8–1).[24] Since the outer portion

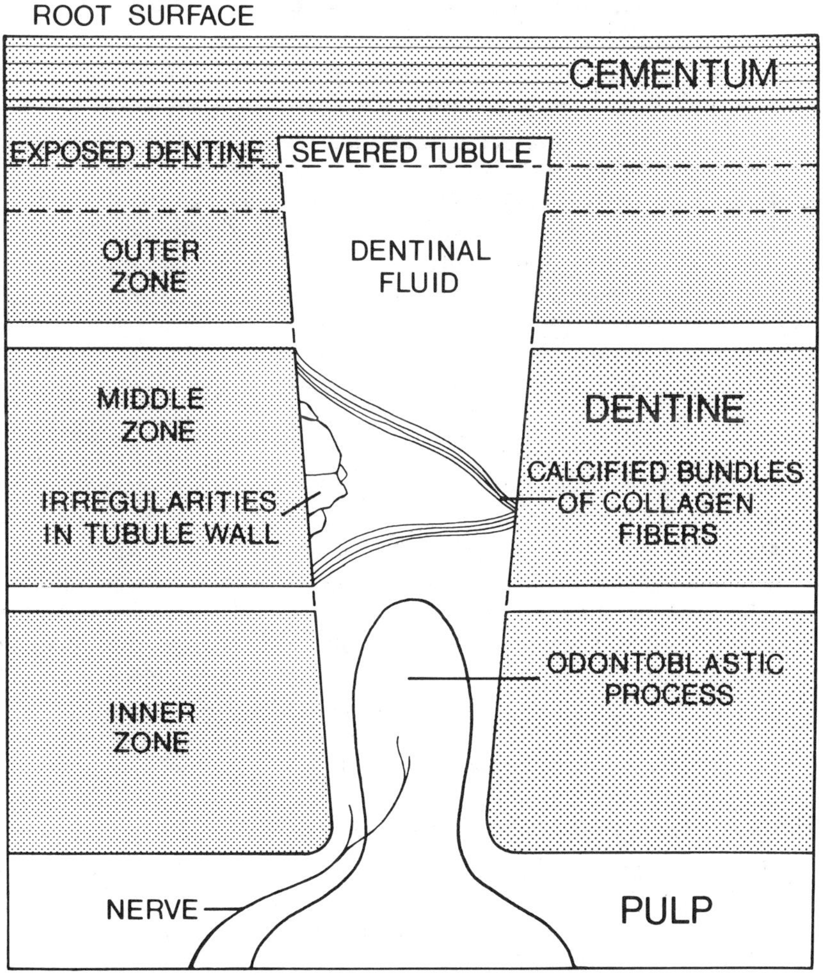

ROOT SURFACE

Fig. 8–1. Diagrammatic representation of a dentinal tubule. When the tubules are severed or exposed, sensitivity occurs, probably from dentinal fluid transmitting pressure changes to mechanoreceptors in the pulp. The odontoblastic process is limited to the inner zone of the dentinal tubules. Irregularities and calcified bundles of collagen fibers may be seen in the middle and outer zone of the tubules. Desensitizing agents, by blocking the opening of the tubules, insulate the dentinal fluid from external stimuli. (Adapted from Pashley.[28])

of the dentin is sensitive to stimuli, a special mechanism must exist to conduct these external stimuli to the nerve endings in the pulp. The most widely accepted theory postulates that these nerves are activated by a *hydrodynamic mechanism*:[25,26] a rapid flow of fluid in the tubules, either inward or outward, resulting in displacement of the tubular contents at the pulpodentinal border. Pain is then produced by distortion of the pulp tissue and activation of mechanoreceptors.[27,28]

Evaluation of Compounds Used to Treat Dentinal Sensitivity

If the hydrodynamic mechanism theory is correct, then to be effective, any desensitizing agent must mechanically block the tubules, either by precipitation of compounds or by surface coating.[27] In addition, a clinically suitable desensitizing agent must be nonirritating to the pulp, relatively painless on application, easily applied, rapid in action, effective for an extended period of time, nonstaining, and consistently effective.[29]

To assess the efficacy of any agent or procedure, an objective clinical means of measuring altered dentinal sensitivity is desirable. Many of the earlier investigations relied on subjective evaluation by patients, but unfortunately such studies "have little scientific basis and belong in the realm of testimonials."[30] More recent clinical studies have used devices whereby the stimulus (electrical or thermal) can be closely controlled and measured.[31-34] With electrical stimulation the threshold of sensitivity is approached slowly, patient discomfort is minimized, and cooperation is more readily obtained. There is no significant correlation, however, between response to electrical and thermal stimuli in the same subject.[34a]

Nevertheless, the individual patient's pain threshold is still used as the end point, and is therefore variable depending on cultural, psychological, and other factors. A further complication in attempting to evaluate desensitizing agents is that placebo treatments produce almost half as much improvement in desensitization as the active ingredients. Changes of such magnitude in the baseline response to placebo agents create statistical problems in evaluating the active agent. Not surprisingly, therefore, some investigators have resorted to animal model systems[35] or in vitro methods such as measuring changes in the hydraulic conductance of dentinal discs.[36,37]

Many different agents have been tested or used clinically to desensitize root surfaces;[38,39] some of these are listed in Table 8–1. A dentifrice containing 5% potassium nitrate has been recognized by the Council on Dental Therapeutics of the ADA as being effective for desensitization.[40] However, when dentifrices of this type were tested in vitro by measuring changes in hydraulic conductance in dentin, they were not significantly different in reducing dentinal permeability than were the placebo dentifrices, which were identical except for the omission of the active ingredient.[37] These observations suggest that the abrasive particles within dentifrices are small enough to penetrate the orifices of dentinal tubules, thereby partially blocking the

Table 8–1. Agents Used for Treating Dentinal Sensitivity

Calcium hydroxide	Sodium citrate and pluronic gel
Calcium sucrosephosphate	Steroids
Formaldehyde	Strontium chloride
Potassium nitrate	Sodium fluoride
Potassium oxalate	Sodium monofluorophosphate
Resins and adhesives	Stannous fluoride

flow of fluid. Potassium oxalate treatment was successful in blocking dentinal tubules, significantly reducing fluid flow[36,37,41] and nerve responses in experimental animals.[35]

By using an animal model for electrophysiological measurement of sensory nerve activity in the pulp, it has been found that depolarization of the sensory nerve membrane in the pulp-dentin complex is an important factor in the mechanism of desensitization. Accordingly, an alternative explanation for the observed clinical efficacy of potassium nitrate and potassium oxalate is based on the presence of high extracellular potassium ion concentration which serves to depolarize the sensory nerve membrane, thereby preventing propagation of action potentials.

Most recently, the Council on Dental Therapeutics[45] has also accepted certain desensitizing dentifrices that contain strontium chloride or sodium citrate and poloxalene gel. This review will focus primarily on the various fluoride compounds and procedures that have been tested.

Studies using Fluoride Compounds for Desensitization

Sodium Fluoride.   Lukomsky[42] was one of the earliest advocates of fluoride therapy for desensitization. He proposed that an isotonic (0.7%) solution of sodium fluoride be applied to root canals, and that a hypertonic solution or paste (31 to 75% NaF) in glycerin be used for sensitive gingival areas. Lukomsky[42] also described the use of a paste consisting of approximately equal portions of NaF, white clay (kaolin), and glycerin for treating sensitive dentin in cavities. Shortly thereafter, such a paste was tried in the United States and was effective in relieving dentinal sensitivity in about 80% of the applications; however, no control group or treatment was used.[43] This paste is still preferred by some practitioners[44] in treating sensitive roots after periodontal surgery. The tooth is dried and the paste burnished with a metal instrument into the sensitive sites and left in position for 2 min. Shorter application (30 seconds) of the paste reduced sensitivity for about 7 days, but after 14 days sensitivity was no longer significantly different from baseline.[33] Because of the extremely high fluoride concentration in such NaF/kaolin/glycerin pastes, their use is restricted to the dental office. This paste has been accepted by the Council on Dental Therapeutics[45] for the treatment of hypersensitive dentin on the cervical areas of teeth.

Iontophoresis has been used to apply sodium fluoride (usually 2%) to sensitive dentin. Iontophoresis is the process of increasing penetration of a desired ionized substance into surface tissue by the aid of a direct electrical current. The sensitive teeth are isolated, preferably with a rubber dam, and 2% NaF soaked on a cotton pellet is applied to the exposed dentin by means of a disposable plastic holder placed over the electrode tip. The electrode (cathode) is connected through a wire lead to the negative pole of a direct current source. The return electrode (anode) is connected to the positive pole of the power unit and strapped to the volar surface of the patient's forearm. The current is adjusted so as not to exceed 1 mA of electricity/min/tooth.[46] Sodium fluoride iontophoresis provided better desensitization than treatment with 33% NaF paste in a double-blind study.[47] In another study, 2% NaF was found equally effective with or without iontophoresis,[48] but this study has been criticized because of wrong polarity (positive) at the tooth electrode. Gangarosa and collaborators[49-52] have advocated F iontophoresis in preference to topical application to provide lasting desensitization. However, these clinical trials have relied on air blast for stimulation and a subjective nonparametric discomfort scale (0 to 4) for evaluation. With one exception, no control procedure was tested. In that single exception, sodium chloride iontophoresis was performed for comparison. However, increased sensitivity resulted which was so distressing to the patients that no other control was attempted.[50] The Council on Dental Therapeutics[40] has not recognized the efficacy of sodium fluoride iontophoresis for desensitization.

Stannous Fluoride. A recent trade publication[53] has proclaimed the "superiority" of stannous fluoride. The implication is that stannous fluoride is superior to acidulated phosphate fluoride and sodium fluoride in plaque inhibition, caries reduction, and desensitization. As regards desensitization, no such direct comparisons have been made; rather, stannous fluoride in glycerin with carboxymethycellulose was found to be significantly better than a *placebo* gel.[54] However, in another study, patients using 0.4% stannous fluoride in glycerin did not exhibit a significant improvement over the control group using a nonfluoride-containing dentifrice (71% vs 60%).[55] More concentrated SnF_2 solutions (2.9%) applied once for 5 min gave a measurable reduction in thermal sensitivity when compared with a water control.[34] A stannous fluoride dentifrice applied using an electroionizing toothbrush with batteries gave greater desensitization than the same dentifrice applied without iontophoresis, but was not significantly better than a strontium chloride dentifrice applied without iontophoresis.[31] In this study a positive charge was supplied by the electroionizing brush with batteries. This suggested that the observed desensitization could have resulted from iontophoresis of the stannous ion, which could form insoluble salts with phosphate in the dentinal tubules, thereby occluding them.

Sodium Monofluorophosphate. Self-application of a dentifrice containing sodium monofluorophosphate as the active anticaries ingredient has also been reported to reduce the sensitivity of dentin.[56-60] Whether this effect is due to fluoride or to other ingredients of the dentifrice entering the orifice of dentinal tubules and blocking them remains to be established.

REFERENCES

1. Daly, T.E., Drane, J.B., and MacComb, W.S.: Management of problems of the teeth and jaws in patients undergoing irradiation. Am. J. Surg., *124*:539–542, 1972.
2. Dreizen, S., Brown, L.R., Daly, T.E., and Drane, J.B.: Prevention of xerostomia-related dental caries in irradiated cancer patients. J. Dent. Res., *56*:99–104, 1976.
3. Johansen, E., and Olsen, T.: Topical fluoride in the prevention and arrest of dental caries. *In* Continuing Evaluation of the Use of Fluorides. Edited by E. Johansen, D.R. Taves, and T.O. Olsen. A.A.S. Selected Symposium 11. Boulder, Colorado, West View Press, 1979, pp. 61–110.
4. Wescott, W.B., Starcke, E.N., and Shannon, I.L.: Chemical protection against post-irradiation dental caries. Oral Surg., *40*:709–719, 1975.
5. Keene, H.J., Fleming, T.J., Brown, L.R., and Dreizen, S.: Lactobacilli and *S. mutans* in cancer patients using fluoride gels. J. Dent. Res. (Special Issue), *63*:281 (Abstract 429), 1984.
6. Englander, H.R., Keyes, P.H., and Gestwicki, M.: Clinical anticaries effect of repeated topical sodium fluoride applications by mouth pieces. J. Am. Dent. Assoc., *75*:638–644, 1967.
7. Brown, L.R., et al.: Microbiological comparisons of carious and noncarious root and enamel tooth surfaces. J. Dent. Res., *62*:295 (Abstract 1137), 1983.
8. Fleming, T.J.: Use of topical fluoride by patients receiving cancer therapy. Curr. Probl. Cancer, *7*:37–41, 1983.
9. Birkeland, J.M., and Torrell, P.: Caries-preventive fluoride mouthrinses. Caries Res., *12* (Suppl. 1):38–51, 1978.
10. Forrester, D.J., and Horowitz, H.S.: Individual topical fluoride therapy. *In* Pediatric Dental Medicine. Edited by D.J. Forrester, M.L. Wagner, and J. Fleming. Philadelphia, Lea & Febiger, 1981, pp. 320–332.

11. Ripa, L.W.: Fluoride rinsing: What dentists should know. J. Am. Dent. Assoc., *102*:477–481, 1981.
12. Bibby, B.G., Zander, H.A., McKelleget, M., and Labunsky, B.: Preliminary reports on the effect on dental caries of the use of sodium fluoride in a prophylactic cleaning mixture and in a mouthrinse. J. Dent. Res., *25*:207–211, 1946.
13. Silverman, S., and Greenspan, O., personal communication.
14. Reed, M.W.: Clinical evaluation of three concentrations of sodium fluoride in dentifrices. J. Am. Dent. Assoc., *87*:1101–1404, 1973.
15. Koch, G., Petersen, L.G., Kling, E., and Kling, L.: Effect of 250 and 1000 ppm fluoride dentifrice on caries. Swed. Dent. J., *6*:233–238, 1982.
16. Swango, P.A.: The use of topical fluorides to prevent dental caries in adults: A review of literature. J. Am. Dent. Assoc., *107*:447–450, 1983.
17. Billings, R.J., Brown, L.R., and Kaster, A.G.: In vivo studies on incipient and shallow root caries. J. Dent. Res. (Special Issue), *63*:257 (Abstract 777), 1984.
18. Al-Joburi, W., Legler, D., and Jamison, H.: Root caries: Control of lesions by iodine-fluoride therapy. J. Dent. Res. (Special Issue), *61*:340 (Abstract 1459), 1982.
19. Stamm, J.W., and Banting, D.W.: Comparison of root caries prevalence in adults with lifelong residence in fluoridated and non-fluoridated communities. J. Dent. Res. (Special Issue A), *59*:405 (Abstract 552), 1980.
20. Rotilie, J.A., McDaniel, T., and Rosen, S.: Root surface caries in the molar teeth of rice rats. III. Inhibition of root surface caries by fluoride. J. Dent. Res., *56*:1498, 1977.
21. Rosen, S., Beck, F.M., and Beck, E.X.: Effect of sodium fluoride dentifrice on root surface caries. J. Dent. Res. (Special Issue), *63*:238 (Abstract 609), 1984.
22. Byers, M.R., and Kish, S.J.: Delineation of somatic nerve endings in rat teeth by radioautography of axon-transported protein. J. Dent. Res., *55*:419–425, 1976.
23. Lilja, J.: Sensory differences between crown and root dentin in human teeth. Acta Odontol. Scand., *38*:285–291, 1980.
24. Brannstrom, M., and Garberoglio, R.: The dentinal tubules and the odontoblast processes, a scanning electron microscopic study. Acta Odontol. Scand., *30*:291–311, 1972.
25. Brannstrom, M.: Sensitivity of dentine. Oral Surg., *21*:517–526, 1966.
26. Brannstrom, M., Linden, L.A., and Astrom, A.: The hydrodynamics of the dental tubule and of pulp fluid. A discussion of its significance in relation to dentinal sensitivity. Caries Res., *1*:310–317, 1967.
27. Dowell, P., and Addy, M.: Dentine hypersensitivity—a review: Etiology, symptoms and theories of pain production. J. Clin. Periodontol., *10*:341–350, 1983.
28. Pashley, D.H.: Dentin conditions and diseases. *In* CRC Handbook of Experimental Aspects of Oral Biochemistry. Edited by E.P. Lazzari. Boca Raton, Florida, CRC Press, Inc., 1983, pp. 97–119.
29. Grossman, L.E.: The treatment of hypersensitive dentine. J. Am. Dent. Assoc., *22*:592–602, 1935.
30. Everett, F.G., Hall, W.B., and Phatak, N.M.: Treatment of hypersensitive dentine. J. Oral Ther. Pharmacol., *2*:300–310, 1966.
31. Johnson, R.H., Zulgar-Nain, B.J., and Koval, J.J.: The effectiveness of an electro-ionizing toothbrush in the control of dentinal hypersensitivity. J. Periodontol., *53*:353–359, 1982.
32. Stark, M.M., and Pelzner, R.: Measurement of dentinal hypersensitivity. Compend. Continuing Educ. Dent. (Suppl. 3), 1982, pp. 105–107.
33. Tarbet, W.J., Silverman, G., Stolman, J.M., and Fratarcangelo, P.A.: An evaluation of two methods for quantitation of dentinal hypersensitivity. J. Am. Dent. Assoc., *98*:914–918, 1979.
34. Thrash, W.J., Dorman, H.L., and Smith, F.D.: A method to measure pain associated with hypersensitive dentin. J. Periodontol., *54*:160–162, 1983.
34a. Thrash, W.J., Blong, M.A., Volding, B.L., and Jones, D.L.: The relationship of electrical to thermal stimulation in pain research. J. Dent. Res., *63*:272 (Abstract 910), 1984.
35. Narhi, M., Hirvonen, T., and Huopaniemi, T.: Sensitivity of dentine. Acupuncture Electro-Ther. Res. Int. J., *8*:143–148, 1983.
36. Greenhill, J.D., and Pashley, D.H.: The effects of desensitizing agents on the hydraulic conductance of human dentine in vitro. J. Dent. Res., *60*:686–698, 1981.
37. Pashley, D.H., et al.: Dentin permeability: Effect of desensitizing dentifrices in vitro. J. Periodontol, 1984, (in press).
38. Addy, M., and Dowell, P.: Dentine hypersensitivity—a review: Clinical and in vitro evaluation of treatment agents. J. Clin. Periodontol., *10*:351–363, 1983.
39. Peden, J.W.: Dental hypersensitivity. J. West. Soc. Periodontol., *25*:75–83, 1977.
40. Council on Dental Therapeutics. Evaluation of Denquel sensitive teeth toothpaste. J. Am. Dent. Assoc., *105*:80, 1982.
41. Hirvonen, T., Narhi, M., and Huopaniemi, T.: A SEM-replica and neurophysiological study on mechanisms of dentine sensitivity. J. Dent. Res., *63*:574 (Abstract 22), 1984.

42. Lukomsky, E.H.: Fluorine therapy for exposed dentin and alveolar atrophy. J. Dent. Res., *20*:649–659, 1941.
43. Hoyt, W.H., and Bibby, B.G.: Use of sodium fluoride for desensitizing dentine. J. Am. Dent. Assoc., *30*:1372–1376, 1943.
44. Carranza, F.A.: Treatment of sensitive roots. *In* Glickman's Clinical Periodontology. 6th Ed. Philadelphia, W.B. Saunders Co., 1984, pp. 769–770.
45. Council on Dental Therapeutics. Categorical Listing of Accepted Dental Products. Chicago, American Dental Association, February 23, 1983, p. 10.
46. Gangarosa, L.P.: Iontophoresis in Dental Practice. Chicago, Quintessence, 1983.
47. Murthy, K.S., Talim, S.T., and Singh, I.: A comparative evaluation of topical application and iontophoresis of sodium fluoride for desensitization of hypersensitive dentine. Oral Surg., *36*:448–458, 1973.
48. Minkov, G., Marami, I., Gedalia, I., and Garfunkel, A.: The effectiveness of sodium fluoride treatment with and without iontophoresis on the reduction of hypersensitive dentine. J. Periodontol., *46*:246–249, 1975.
49. Gangarosa, L.P., et al.: Desensitizing hypersensitive dentin by iontophoresis with fluoride. NY State Dent. J., *44*:92–94, 1978.
50. Gangarosa, L.P., and Park, N.H.: Practical considerations in iontophoresis of fluoride for desensitizing dentin. J. Prosthet. Dent., *39*:173–178, 1978.
51. Gangarosa, L.P., and Heuer, G.A.: A practical technique for treating tooth hypersensitivity. Dent. Survey, *55*:37–40, 1979.
52. Gangarosa, L.P.: Iontophoretic application of fluoride by tray techniques for desensitization of multiple teeth. J. Am. Dent. Assoc., *102*:50–52, 1981.
53. Anonymous. Stannous fluoride as a plaque inhibitor. Human clinical studies. Gel-Kam Prevent. Dent. Rev., *4* (2). Dallas, Texas, Scherer Labs Inc. (no date).
54. Miller, J.T., Shannon, I., Kilgore, W., and Bookman, J.: Use of a water-free stannous fluoride containing gel in the control of dentinal hypersensitivity. J. Periodontol., *40*:490–491, 1969.
55. Zinner, D.D., Duany, L.F., and Lutz, H.J.: A new desensitizing dentifrice: Preliminary Report. J. Am. Dent. Assoc., *95*:982–985, 1977.
56. Bolden, T.E., Volpe, A.R., and King, W.J.: The desensitizing effect of a sodium monofluorophosphate dentifrice. Periodontics, *6*:112–114, 1968.
57. Hazen, S.P., Volpe, A.R., and King, W.J.: Comparative desensitizing effect of dentifrices containing sodium monofluorophosphate, stannous fluoride and formalin. Periodontics, *6*:230–232, 1968.
58. Hernandez, F., et al.: Clinical study evaluating the desensitizing effect and duration of two commercially available dentifrices. J. Periodontol., *43*:367–372, 1972.
59. Kanouse, M.C., and Ash, M.M.: The effectiveness of sodium monofluorophosphate dentifrice on dental hypersensitivity. J. Periodontol., *40*:38–40, 1969.
60. Shapiro, W.B., Kaslick, R.S., Chasens, A.I., and Weinstein, D.: Controlled clinical comparison between a strontium chloride and sodium monofluorophosphate toothpaste in diminishing root hypersensitivity. J. Periodontol., *41*:523–525, 1970.

Section III

HARRY BOHANNAN, MODERATOR

Chapter 9
ARE ALL FLUORIDE DENTIFRICES THE SAME?

George K. Stookey

Beginning with the first clinical evaluation of a fluoride-containing dentifrice in the 1940s, a literature review of the numerous reports on clinical investigations of fluoride (F) dentifrices has been performed. The first fluoride-containing dentifrice, which was demonstrated to have significant cariostatic activity, contained stannous fluoride (SnF_2) with calcium pyrophosphate ($Ca_2P_2O_7$) as the abrasive or polishing agent. Subsequent studies revealed similar cariostatic benefits using stannous fluoride with other abrasive systems. Sodium monofluorophosphate (Na_2PO_3F) formulations were proven to measurably reduce incremental caries rates in a series of studies conducted during the 1960s. Later investigations also demonstrated the cariostatic efficacy of sodium monofluorophosphate when used with a variety of different abrasive systems. Similarly, the use of dentifrices containing sodium fluoride (NaF) with several different abrasive systems has been shown to reduce the incidence of dental caries. There can be no doubt that fluoride-containing dentifrices offer a practical means for contributing to the control of dental caries.

In order to respond to the question "Are all fluoride dentifrices the same?" it seems appropriate, if not necessary, to review the past as a means of positioning the current thinking and the present state of the art. A review of the literature revealed more than 140 clinical studies with fluoride dentifrices and undoubtedly a few additional studies, probably published in foreign journals, have been inadvertently missed.

REVIEW OF THE LITERATURE

By the early 1940s it had been established that the presence of fluoride in the drinking water dramatically decreased the prevalence of dental caries. Although the mechanism of action was not clearly known, it was thought to be primarily systemic, whereby the benefit was due to the incorporation of fluoride into developing enamel resulting in the formation of a fluorapatite structure, which was more resistant to acids. Scientists reasoned that this same effect might be achieved by the simple application of fluoride to the tooth surfaces. They also speculated that the use of treatment modalities having higher concentrations of fluoride with reduced exposure periods might be a means of compensating for prolonged exposure to very low levels of fluoride.

In 1942, Bibby[1] initiated the first clinical evaluation of dentifrices containing fluoride. Similar investigations were reported during the next decade (Table 9–1). In

Table 9–1. Early NaF Dentifrice Clinical Studies

Study	Abrasive System	Result
Bibby (1945)[1]	$CaHPO_4$	Not Significant
Bibby & Wellock (1948)[2]	$CaHPO_4$	Not Significant
Wellock & Bibby (1949)[3]	$CaHPO_4$	Not Significant
Winkler et al. (1953)[4]	$CaCO_3$	Not Significant
Muhler et al. (1955)[5]	$Ca_2P_2O_7$	Not Significant
Muhler (1957)[6]	$Ca_2P_2O_7$	Not Significant
Kyes et al. (1961)[7]	$(NaPO_3)_x/CaHPO_4$	Not Significant
Brudevold & Chilton (1966)[8]	$CaHPO_4$	Not Significant

these early studies the fluoride was provided as sodium fluoride and, for all practical purposes, was simply incorporated into existing dentifrice formulations. None of these early formulations, however, resulted in a significant reduction in the incidence of dental caries. Throughout the present review, significance has been considered as a probability of 0.05 or less on the basis of DMFS increments. The determination of a significant benefit was made by direct comparison of caries increments in two or more groups of subjects, one of which was provided a nonfluoride or placebo dentifrice.

In a 1967 review of sodium fluoride dentifrices, Gron and Brudevold[9] noted that the probable reasons for the failure of the early studies were the incompatibility of the added fluoride with the dentifrice constituents, particularly the abrasive system, and inadequacies in study design, most notably, the absence of supervised toothbrushing in an era when toothbrushing was not as much of a daily routine for children as it is today.

Before continuing, let me point out that the term abrasive has been commonly used to refer to the cleaning and polishing agent present in a dentifrice. Numerous clinical studies have repeatedly shown that this component of the dentifrice is absolutely essential for the removal and control of stained pellicle.[10–16] The use of the term abrasive implies that the material may abrade dental hard tissues and therefore be detrimental. While it is possible to detect hard tissue abrasion with sophisticated procedures,[17] two long-term clinical studies[18,19] with conventional dentifrice abrasive systems failed to demonstrate clinical abrasion to enamel or restorations. It has been shown that abrasion of exposed root surfaces is more dependent upon the manner of toothbrushing than the abrasivity of the dentifrice.[20] There is no need for undue concern for abrasivity with conventional dentifrices.

If we examine the nature of the dentifrice abrasives used in these early studies, it is apparent that they all contained calcium provided as various calcium phosphates, calcium carbonate or a blend of dicalcium phosphate and insoluble sodium metaphosphate. As was noted by Ericsson[21] as well as other investigators, sodium fluoride readily reacts with calcium carbonate and other calcium-containing abrasives to form calcium fluoride. This formation results in the existence of the fluoride in a chemical form which is no longer reactive with enamel. Thus, the lack of available and reactive ionic fluoride in the early fluoride dentifrices was most likely the major reason for their failure to prevent caries.

In 1952, a clinical study was initiated with a different dentifrice system containing stannous fluoride in combination with a calcium phosphate which had been heat-

Table 9–2.   Clinical Studies with SnF_2-$Ca_2P_2O_7$ Dentifrice

Total of 45 Clinical Trials Reported	
Significantly Less Caries	36 Studies
No Significant Benefit	9 Studies

treated to increase fluoride compatibility. Two years later, in 1954, the first report of a clinical decrease in the incidence of caries with a fluoride-containing dentifrice, as compared to the similar use of a nonfluoride dentifrice, was published.[22] Similar investigations were reported during the subsequent decade that served to verify the cariostatic effectiveness of this formulation (Table 9–2).

It is also worthy to note that the Council on Dental Therapeutics of the American Dental Association (ADA) decided to classify this dentifrice in category B in 1960,[66] and classified it in category A, indicative of acceptance as a therapeutic measure, in 1964[67] upon completion of additional investigations. This classification provided the necessary impetus for industrial organizations to invest countless millions of dollars in research, which has continued to this day to the betterment of the general public, the dental profession, the ADA, and dentifrice manufacturers.

The original stannous fluoride-calcium pyrophosphate formulation was essentially unchanged through 1981. During those 25 years, it was the subject of more than 40 reported clinical trials which verified its efficacy[5,7,8,23–65] and resulted in its use as a standard of reference for the identification of comparable or improved fluoride dentifrice formulations (Table 9–2).

Once an effective caries-preventive composition had been identified and the potential value of such an approach for the partial control of caries had been recognized, a significant amount of research was undertaken in an attempt to develop even more effective dentifrice formulations. In general, these investigations were directed toward two approaches: (1) the use of different fluoride systems, and (2) the use of abrasive compositions that would yield greater amounts of available fluoride for reaction with enamel. These investigations resulted in numerous clinical trials that were reported during the 1960s.

Since stannous fluoride had been shown to possess cariostatic activity and it was thought that the amount of available fluoride in a dentifrice was a key factor, efforts were made to identify abrasive systems that were more compatible with stannous fluoride. One such abrasive system was composed of insoluble sodium metaphosphate (IMP) with a small amount, typically about 5%, of dicalcium phosphate (DCP). The results of a number of clinical investigations conducted during the 60s clearly indicated that the use of stannous fluoride dentifrices formulated with this abrasive system resulted in a significant decrease in the incidence of dental caries when compared to the use of a nonfluoride dentifrice (Table 9–3). Indeed, the results of these studies led to the provisional acceptance, or B classification, of two additional dentifrices, which became available at that time.[73,74]

Included in some of these clinical trials[33,34,40] were direct comparisons of stannous fluoride dentifrices formulated with either calcium pyrophosphate or the insoluble sodium metaphosphate-dicalcium phosphate abrasive system. The results of these comparative tests demonstrated a similar amount of benefit from the two dentifrices (Table 9–4).

Another series of clinical tests was conducted using stannous fluoride formulations

Table 9–3. Clinical Studies with SnF_2-IMP/DCP Dentifrices

Study Reported	Significant Benefit
Henriques et al. (1964)[33]	Yes
Mergele et al. (1964)[34]	Yes
Slack & Martin (1964)[68]	No
Thomas & Jamison (1966)[40]	Yes
Fullmer et al. (1966)[69]	Yes
Segal et al. (1967)[70]	Yes
Naylor & Emslie (1967)[71,72]	Yes

Table 9–4. Clinical Studies Comparing SnF_2-$Ca_2P_2O_7$ and SnF_2-$(NaPO_3)_x$/$CaHPO_4$ Dentifrices

Study Reported	Significant Difference
Henriques et al. (1964)[33]	No
Mergele et al. (1964)[34]	No
Thomas & Jamison (1966)[40]	No

that were free of calcium while still utilizing insoluble sodium metaphosphate as the abrasive. In three of the six clinical investigations, a significant decrease in caries incidence was observed when compared to the results obtained with the use of a nonfluoride dentifrice (Table 9–5). On the basis of these studies, a stannous fluoride-insoluble sodium metaphosphate dentifrice was given provisional acceptance by the ADA and was marketed for a period of time.[78] In two of these studies[8,56] a direct comparison was made between the stannous fluoride-calcium pyrophosphate and stannous fluoride-insoluble sodium metaphosphate products (Table 9–6). Published details of another study comparing these dentifrices were too inadequate to include

Table 9–5. Clinical Studies with SnF_2-IMP Dentifrices

Study Reported	Significant Benefit
Brudevold & Chilton (1966)[8]	Yes
Slack et al. (1967)[75]	No
Fanning et al. (1968)[76]	Yes
Mergele (1968)[77]	No
Frankl & Alman (1968)[51]	No
Slack et al. (1971)[56]	Yes

Table 9–6. Clinical Studies Comparing SnF_2-$Ca_2P_2O_7$ and SnF_2-$(NaPO_3)_x$ Dentifrices

Study Reported	Significant Difference
Brudevold & Chilton (1966)[8]	Yes
Slack et al. (1971)[56]	No

in this review.[79] In both of the cited studies a numerically greater benefit from the insoluble sodium metaphosphate formulation was observed. This difference was also significant in one of the tests.

More recently, two studies[62,64] were reported in which a highly compatible silica abrasive system was utilized in stannous fluoride dentifrices (Table 9–7). In both investigations, statistically significant decreases in caries increments were observed when compared to the similar use of a nonfluoride dentifrice. On the basis of these studies a stannous fluoride-silica dentifrice was accepted by the ADA and marketed for several years.[80]

Again, both of these tests included a direct comparison with the stannous fluoride-calcium pyrophosphate formulation (Table 9–8). In spite of apparent differences in the amount of available fluoride in these two formulations, similar levels of cariostatic activity for the two products were observed in both investigations. It is apparent that concerted efforts were made by several teams of scientists to develop stannous fluoride dentifrices with increased concentrations of available fluoride through the use of more compatible abrasive systems, but, with only one exception[8] these efforts failed to provide a significant increase in the cariostatic benefits of this compound.

The alternative approach to identify dentifrice compositions having greater cariostatic potential than stannous fluoride involved a search for more effective compounds. While most of these efforts were ultimately directed toward sodium monofluorophosphate and sodium fluoride systems, it should be noted that several other compounds were also investigated. For example, promising results were reported following the topical application of other tin-containing compounds such as stannous chlorofluoride,[81] potassium trifluorostannite,[82] and stannous hexafluorozirconate.[83] The utility of these compounds in dentifrices, however, has not been reported.

As shown in Table 9–9, one group of investigators obtained significant caries reductions with mixtures of sodium and stannous fluoride.[84,85] Another dentifrice trial utilized ammonium fluoride and obtained promising results.[86] Gerdin and Koch conducted dentifrice studies[87–89] involving the use of a mixture of potassium fluoride and manganese chloride and reported this system to be superior to sodium fluoride; additional studies with this system appear to be in order.

Marthaler has reported (Table 9–10) significant benefits from the use of amine

Table 9–7. Clinical Studies with SnF_2-Silica Dentifrices

Study Reported	Significant Benefit
Fogels et al. (1979)[62]	Yes
Abrams & Chambers (1980)[64]	Yes

Table 9–8. Clinical Studies Comparing SnF_2-$Ca_2P_2O_7$ and SnF_2-Silica Dentifrices

Study Reported	Significant Difference
Fogels et al. (1979)[62]	No
Abrams & Chambers (1980)[64]	No

Table 9–9. Clinical Dentifrice Studies Using Lesser Known Fluoride Compounds or Systems

Study Reported	Fluoride System	Significant Benefit
Held & Spirgi (1965)[84]	NaF/SnF$_2$	Yes
Held & Spirgi (1968)[85]	NaF/SnF$_2$	Yes
Geiger et al. (1971)[86]	NH$_4$F	Yes
Gerdin (1972)[87]	KF/MnCl$_2$	Yes
Koch (1972)[88]	KF/MnCl$_2$	Yes
Gerdin (1974)[89]	KF/MnCl$_2$	Yes

Table 9–10. Clinical Studies with Amine Fluoride Dentifrices

Study Reported	Significant Benefit
Marthaler (1968)[90]	Yes
Patz & Naujoks (1970)[91]	Yes
Marthaler (1974)[92]	Yes
Kunzel et al. (1977)[93]	Yes
Ringelberg et al. (1979)[61]	Yes
Cahen et al. (1982)[94]	Yes

fluoride dentifrices that used a mixture of two different amine fluorides,[90,92] and similar results were observed by Patz and Naujoks.[91] Although these observations were generally confirmed by Ringelberg et al.,[61] the level of efficacy was comparable to that obtained with a stannous fluoride dentifrice. On the other hand, Cahen and coworkers[94] found the amine fluoride dentifrice to be significantly more effective than a dentifrice containing an elevated concentration of sodium monofluorophosphate. None of these alternative fluoride compounds has been either adequately evaluated in dentifrice systems or has appeared to be sufficiently promising as compared to more conventional agents to justify further consideration at this time.

The first report of a clinical trial involving the use of sodium monofluorophosphate in a dentifrice was by Finn and Jamison[32] (Table 9–11). The results of this trial suggested that this compound might be more effective than stannous fluoride.[32] Numerous investigations with this fluoride compound have since been reported. Early studies with sodium monofluorophosphate used dentifrices formulated with either insoluble sodium metaphosphate alone or with a small amount of added calcium

Table 9–11. Clinical Studies with Sodium Monofluorophosphate-Insoluble Sodium Metaphosphate Dentifrices

Study Reported	Significant Benefit
Finn & Jamison (1963)[32]	Yes
Fanning et al. (1968)[76]	Yes
Mergele (1968)[77]	Yes
Mergele (1968)[49]	Yes
Kinkel & Stolte (1968)[95]	Yes
Patz & Naujoks (1969)[96]	No
Barlage et al. (1981)[97]	Yes
Cahen et al. (1982)[94]	Yes

phosphate. The results of the studies that utilized insoluble sodium metaphosphate alone clearly indicated this dentifrice reduced the incidence of dental caries as compared to the similar use of a nonfluoride dentifrice.

Several additional studies utilized sodium monofluorophosphate with insoluble sodium metaphosphate containing a small amount of dicalcium phosphate (Table 9–12). Again, significant decreases in the incidence of caries were observed in most of the investigations with the magnitude of the benefit generally comparable to that observed with calcium-free insoluble sodium metaphosphate formulations.

These observations of the effectiveness of sodium monofluorophosphate in both calcium-free and calcium-containing dentifrices prompted a significant number of studies using dentifrices formulated with a variety of abrasive systems. Several studies were reported in which dicalcium phosphate was used as the polishing agent, and all of these studies demonstrated significant cariostatic benefits (Table 9–13).

Another investigation utilized calcium pyrophosphate as the abrasive system (Table 9–14). This formulation was also shown to significantly decrease the incidence of dental caries when compared to a nonfluoride dentifrice.[58]

Somewhat more recently studies were conducted using hydrated alumina as the

Table 9–12. Clinical Studies with Sodium Monofluorophosphate-Insoluble Sodium Metaphosphate/Dicalcium Phosphate Dentifrices

Study Reported	Significant Benefit
Møller et al. (1968)[98]	Yes
Thomas & Jamison (1970)[99]	Yes
Peterson (1979)[100]	No
Glass et al. (1983)[101]	Yes
Blinkhorn et al. (1983)[102]	Yes
Ashley et al. (1977)[103]	Yes

Table 9–13. Clinical Dentifrice Studies with Sodium Monofluorophosphate-Dicalcium Phosphate Formulations

Study Reported	Significant Benefit
Naylor et al. (1967)[71,72]	Yes
Takeuchi et al. (1968)[104]	Yes
Onisi & Tani (1970)[55]	Yes
Kinkel & Raich (1974)[105]	Yes
Kinkel et al. (1974)[106]	Yes
Niwa et al. (1975)[107]	Yes
Rijnbeek & Weststrate (1976)[108]	Yes
Kinkel et al. (1977)[109]	Yes

Table 9–14. Clinical Study with Sodium Monofluorophosphate-Calcium Pyrophosphate Dentifrice

Study Reported	Significant Benefit
Zacherl (1972)[58]	Yes

abrasive system (Table 9–15). Again, significant benefits were observed in most of these studies and the lone exception[111] demonstrated significance for DMFT but not for DMFS.

Many studies have also been conducted using sodium monofluorophosphate in calcium carbonate dentifrices (Table 9–16). Significant cariostatic benefits, as compared to the similar use of nonfluoride dentifrices, were observed in all of these studies. Similarly, several investigations have been reported in which this fluoride compound was used with a blend of calcium carbonate and silica (Table 9–17). Again, most of these studies resulted in significant reductions in the incidence of caries.

Finally, three studies have been reported in which silica was used as the abrasive system (Table 9–18) and a significant cariostatic benefit was observed in two of the three investigations. In addition to these investigations, several studies in which significant benefits were reported with sodium monofluorophosphate dentifrices have not been listed since the nature of the abrasive system was either not reported or the necessary details were not available.[124–127]

From this extensive series of studies, which involved a wide variety of abrasive systems, it is quite apparent that sodium monofluorophosphate is compatible with a wide variety of dentifrice polishing agents. Indeed, these studies resulted in the acceptance of several different dentifrice formulations containing sodium monofluor-

Table 9–15. Clinical Studies with Sodium Monofluorophosphate-Hydrated Alumina Dentifrices

Study Reported	Significant Benefit
Andlaw & Tucker (1975)[110]	Yes
Hodge et al. (1980)[111]	No
Murray & Shaw (1980)[112]	Yes
Andlaw et al. (1983)[113]	Yes

Table 9–16. Clinical Studies with Sodium Monofluorophosphate-Calcium Carbonate Dentifrices

Study Reported	Significant Benefit
Torell & Ericsson (1965)[35]	Yes
Torell (1969)[114]	Yes
Peterson & Williamson (1975)[115]	Yes
Mainwaring & Naylor (1978)[116]	Yes
Glass & Shiere (1978)[117]	Yes
Peterson (1979)[100]	Yes
Mainwaring & Naylor (1983)[118]	Yes

Table 9–17. Clinical Studies with Sodium Monofluorophosphate-Calcium Carbonate/Silica Dentifrices

Study Reported	Significant Benefit
Forsman (1974)[119]	No
Naylor & Glass (1979)[120]	Yes
Glass (1981)[121]	Yes
Glass et al. (1983)[101]	Yes

Table 9–18. Clinical Studies with Sodium Monofluorophosphate-Silica Dentifrices

Study Reported	Significant Benefit
Forsman (1974)[119]	No
Howat et al. (1978)[122]	Yes
Rule et al. (1982)[123]	Yes

ophosphate as therapeutic compositions by the Council on Dental Therapeutics of the ADA.[128-131] There have been very few direct comparisons of sodium monofluorophosphate formulations having different abrasive systems in the same study;[100,101,119,132] therefore, there are few data to suggest a preferred abrasive system for use with this fluoride compound.

As noted earlier it was initially suggested that sodium monofluorophosphate was a more effective fluoride agent than stannous fluoride. This suggestion was based upon the initial study reported by Finn and Jamison.[32] As of this date, there have been a total of eight clinical trials in which a sodium monofluorophosphate formulation was directly compared with a stannous fluoride formulation (Table 9–19). In five of these studies, a stannous fluoride-calcium pyrophosphate system was compared. Numerical differences favored sodium monofluorophosphate in six studies and stannous fluoride in the other two tests. No significant difference between the two fluoride compositions was observed in six of the eight studies. The two exceptions have been studied by other reviewers and it was found that "Although the data in two of these studies [i.e., the Finn and Jamison[32] and the Frankl and Alman[51] studies] showed that groups using the monofluorophosphate product had fewer new carious lesions, the studies do not permit a definitive evaluation since they contain inadequate controls."[133] If one accepts this statement from the Council and notes that none of the other six studies observed a significant difference between sodium monofluorophosphate and stannous fluoride compositions, then one must conclude that the clinical data indicate a comparable level of cariostatic activity of dentifrices containing these two fluoride systems.

Quite recently, several additional approaches designed to increase the cariostatic activity of sodium monofluorophosphate dentifrices have been investigated. These approaches have included: (1) the use of an additive, calcium glycerophosphate; (2)

Table 9–19. Clinical Studies Comparing Stannous Fluoride and Sodium Monofluorophosphate Dentifrices

Study Reported	Abrasive System		Significant Difference
	SnF_2	Na_2PO_3F	
Finn & Jamison (1963)[32]	$Ca_2P_2O_7$	$NaPO_3$	Yes
Naylor et al. (1967)[71,72]	$NaPO_3/CaHPO_4$	$CaHPO_4$	No
Mergele (1968)[49]	$Ca_2P_2O_7$	$NaPO_3/CaHPO_4$	No
Frankl & Alman (1968)[51]	$Ca_2P_2O_7$	$NaPO_3$	Yes
Fanning et al. (1968)[76]	$NaPO_3$	$NaPO_3$	No
Zacherl (1972)[58]	$Ca_2P_2O_7$	$Ca_2P_2O_7$	No
Mergele (1968)[77]	$NaPO_3$	$NaPO_3/CaHPO_4$	No
Onisi & Tani (1970)[55]	$Ca_2P_2O_7$	$CaHPO_4$	No

the use of increased concentrations of sodium monofluorophosphate; and (3) the use of mixtures of sodium monofluorophosphate with sodium fluoride.

The potential value of the addition of calcium glycerophosphate to enhance the cariostatic activity of dentifrices containing sodium monofluorophosphate has been evaluated in three clinical trials, each of which included a conventional sodium monofluorophosphate formulation as a positive control (Table 9–20). In none of these studies did the addition of calcium glycerophosphate significantly alter the cariostatic activity of the fluoride system, although the difference approached significance in one instance.[118] On the basis of these reports, it does not appear that the inclusion of this additive significantly enhances the cariostatic activity of sodium monofluorophosphate.

Investigations of different concentrations of sodium monofluorophosphate have actually been pursued from two directions: (1) to determine if lower concentrations were equally effective; and, (2) to determine if higher concentrations resulted in greater benefits (Table 9–21). Two studies compared the use of a lesser concentration of 0.2% with the conventional level of 0.8%. In the study by Forsman,[119] no effect was observed when decreasing the concentration from 0.8 to 0.2%; however, the results of this study must be interpreted with caution since neither concentration of sodium monofluorophosphate provided a significant benefit. In contrast, Mitropoulos and coworkers[134] reported significantly less benefit with dentifrices containing the lesser concentration. When increasing concentrations of sodium monofluorophosphate were used, Barlage and coworkers[97] reported a significant increase in cariostatic activity when the level of fluoride was increased from 0.8% (or 1000 ppm F) to 1.2%, or about 1500 ppm.

Two recent studies[94,135] using the increased fluoride concentration of 1.2% similarly reported significant benefits but no comparison was made to formulations containing lesser fluoride levels (Table 9–22). Three clinical trials[136–138] have used sodium monofluorophosphate dentifrices with markedly increased fluoride concentrations

Table 9–20. Clinical Studies of Calcium Glycerophosphate (CaGP) Added to Sodium Monofluorophosphate Dentifrices

Study Reported	Cariostatic Benefit		Significant Difference
	Na_2PO_3F	Na_2PO_3F/CaGP	
Naylor & Glass (1979)[120]	Yes	Yes	No
Glass et al. (1983)[101]	Yes	Yes	No
Mainwaring & Naylor (1983)[118]	Yes	Yes	No

Table 9–21. Clinical Dentifrice Studies Comparing Different Concentrations of Sodium Monofluorophosphate

Study Reported	Na_2PO_3F Concentrations			Significant Difference
	0.2%	0.8%	1.2%	
Forsman (1974)[119]	X	X		No
Mitropoulos et al. (1982)[134]	X	X		Yes
Barlage et al. (1980)[97]		X	X	Yes

Table 9–22. Clinical Studies with Dentifrices Containing Elevated Concentrations of Sodium Monofluorophosphate

Study Reported	Concentration Studied	Significant Benefit
Cahen et al. (1982)[94]	1.2%	Yes
Hanachowicz (1984)[135]	1.2%	Yes
Hargreaves & Chester (1973)[136]	2.0%	Yes
Lind et al. (1974)[137]	2.0%	Yes
James et al. (1977)[138]	2.0%	Yes

of about 2500 ppm or about 2.5 times the conventional level. In each instance significant benefits were observed; unfortunately, none of these studies reported the use of the conventional fluoride concentration. Therefore, it cannot be determined if the increased concentration of sodium monofluorophosphate resulted in benefits greater than those obtained with conventional levels.

A recent study by Hodge and coworkers[111] investigated the use of elevated concentrations of fluoride obtained by adding sodium fluoride to the conventional level of sodium monofluorophosphate (Table 9–23). In this study a very modest effect, significant only for DMFT, was observed with the conventional concentration of sodium monofluorophosphate. The use of two different formulations containing added sodium fluoride, however, resulted in significantly increased cariostatic benefits. This observation is interesting, but appropriately designed confirmatory studies are lacking. The true efficacy, and particularly any improvement in benefits, derived from increased concentrations of fluoride provided as mixtures of compounds can only be determined by direct comparison with similar concentrations of fluoride provided by the individual compounds. Related to this report are the findings of Mainwaring and Naylor,[118] who noted that the replacement of half the conventional amount of the sodium monofluorophosphate with sodium fluoride resulted in a numerically, but not significantly, greater benefit in spite of the use of an abrasive which is only modestly compatible with sodium fluoride.[21]

The early failures with sodium fluoride dentifrices apparently dampened any enthusiasm for further work for a number of years. In 1965, however, Torell and Ericsson[35] reported a significant reduction in caries with a sodium fluoride dentifrice that utilized a compatible material, sodium bicarbonate, as the polishing agent. During the same year Jiraskova and coworkers[139] also reported that a Czechoslovakian dentifrice containing sodium fluoride resulted in a significant decrease in the incidence of dental caries as compared to the use of a nonfluoride dentifrice.

In a 1967 review of sodium fluoride dentifrices Grøn and Brudevold[9] noted that, while the use of incompatible polishing agents in the early sodium fluoride studies probably accounted largely for their failure, the lower pH of stannous fluoride

Table 9–23. Clinical Dentifrice Study Using Elevated Fluoride Levels Obtained from Mixtures of 0.8% Na_2PO_3F and 0.1% NaF

Study Reported	Cariostatic Benefit		Significant Difference
	Na_2PO_3F	Na_2PO_3F + NaF	
Hodge et al. (1980)[111]	No	Yes	Yes

compositions may have contributed to the success of this agent. These investigators demonstrated that topical applications of sodium fluoride solutions acidulated with soluble phosphate, better known today as APF, significantly reduced incremental caries.[140,141] Therefore, it logically followed that similar systems should be evaluated in dentifrices.

During the next few years the results of four dentifrice studies were reported in which sodium fluoride and soluble phosphate were used as the active system with insoluble sodium metaphosphate as the abrasive (Table 9–24). Two of these studies indicated that this system resulted in a significant decrease in incremental caries. A dentifrice containing this system was marketed for a period of time. Another clinical trial with this fluoride-phosphate system was not included in this review, due to the absence of adequate published details,[79] although a related formulation containing an elevated level of sodium fluoride of 1250 ppm was reported to exert a significant benefit.[142]

During these investigations another series of studies were reported (Table 9–25) in which a particular type of calcium pyrophosphate, identified as a high-beta-phase form, was used as the polishing agent in sodium fluoride formulations. Laboratory tests with this abrasive indicated that it was much more compatible with sodium fluoride than any of the calcium compounds used in the prior investigations with about 60 to 70% of the fluoride being available.[147] The results obtained in six clinical trials with essentially the same sodium fluoride-calcium pyrophosphate system consistently demonstrated significant cariostatic benefits as compared to the use of a nonfluoride dentifrice.[48,143–147] A sodium fluoride dentifrice formulated with this abrasive system was approved by the Food and Drug Administration in 1973.[148]

The study conducted by Zacherl[58] included a direct comparison of sodium fluoride and sodium monofluorophosphate in the calcium pyrophosphate system (Table 9–26). Significant benefits were observed with both compositions and there were no

Table 9–24. Clinical Studies with Sodium Fluoride-Soluble Orthophosphate Dentifrices

Study Reported	Significant Benefit
Brudevold & Chilton (1966)[8]	Yes
Peterson & Williamson (1968)[48]	Yes
Slack et al. (1971)[56]	No
Zacherl (1972)[58]	No

Table 9–25. Clinical Studies with Sodium Fluoride-Modified Calcium Pyrophosphate Dentifrices

Study Reported	Significant Benefit
Reed & King (1970)[143]	Yes
Zacherl (1972)[58]	Yes
Weisenstein & Zacherl (1972)[144]	Yes
Reed (1973)[145]	Yes
Stookey & Beiswanger (1975)[146]	Yes
Ennever et al. (1980)[147]	Yes

Table 9–26. Clinical Comparison of NaF-Modified $Ca_2P_2O_7$ and Na_2PO_3F-$Ca_2P_2O_7$ Dentifrice

Study Reported	Caries Benefit		Significant Difference
	NaF	Na_2PO_3F	
Zacherl (1972)[58]	Yes	Yes	No

differences between the systems, although only about two-thirds of the fluoride from sodium fluoride was available.

During the early 1970s, reports of studies conducted in France[149] and Japan[150] with sodium fluoride dentifrices also appeared. In both instances, significant reductions in caries incidence were reported. In the French study, the effects were observed during a 5-year period in a pedodontic practice. In the study conducted in Japan, a synthetic zeolite (a complex silicate) was used as the abrasive system; this abrasive system also resulted in an intermediate level of fluoride availability.

It is relevant to note that one of the two sodium fluoride formulations evaluated by Ennever and coworkers[147] had an alkaline pH that markedly increased the compatibility such that essentially all of the fluoride was available as ionic fluoride. The clinical efficacy, however, was quite comparable to a neutral formulation with one-third less available fluoride. Similarly, Gerdin[87] observed that an alkaline sodium fluoride system was less effective than a neutral composition. These two observations prove that the amount of available fluoride does not necessarily indicate the bioavailability of the fluoride (or the amount of fluoride which reacts with the dentition) as evidenced by clinical effectiveness.

Concurrent with these latter studies, other investigators studied the effectiveness of sodium fluoride dentifrices that utilized even more compatible abrasive systems (Table 9–27). Of particular interest was the use of polymethylmethacrylate particles as the abrasive in the studies by Koch,[151–153] since this material is extremely compatible with fluoride. Significant cariostatic benefits were observed in all three of these studies.

In a similar effort to utilize sodium fluoride in highly compatible dentifrice compositions, two additional investigations were reported in which silica was used as the polishing agent. In one of these investigations, namely, the Zacherl study,[65] significant decreases in incremental caries as compared to a placebo dentifrice were observed. The other study, conducted by Forsman[119] had been questioned earlier since no significant benefits were observed with either a series of sodium monofluorophosphate or sodium fluoride formulations.

Table 9–27. Clinical Sodium Fluoride Dentifrice Studies with Formulations Having Highly Compatible Abrasive Systems

Study Reported	Abrasive System	Significant Benefit
Koch (1967)[151]	Acrylic	Yes
Koch (1967)[152]	Acrylic	Yes
Koch (1970)[153]	Acrylic	Yes
Forsman (1974)[119]	Silica	No
Zacherl (1981)[65]	Silica	Yes

Two clinical studies[65,154] were conducted in which a direct comparison was made between a neutral sodium fluoride-silica dentifrice and the stannous fluoride-calcium pyrophosphate dentifrice which had served as a point of reference, or positive control, in many previous studies (Table 9–28). In both of these investigations, the sodium fluoride formulation was significantly more effective than the stannous fluoride dentifrice.

Several of the clinical investigations of sodium fluoride dentifrices also included compositions containing sodium monofluorophosphate (Table 9–29). Only the study by Forsman[119] included a placebo, or negative control group. In that study, both fluoride compounds were used in silica abrasive systems and neither resulted in significant reductions in incremental caries. Gerdin[87] compared a sodium fluoride-acrylic dentifrice with a sodium monofluorophosphate-calcium carbonate product and found the neutral sodium fluoride composition significantly more effective. Edlund and Koch[155] similarly compared dentifrices containing these two fluoride systems and observed significantly greater benefits with the sodium fluoride formulation. Edward and Torell[156] also compared four different sodium fluoride and sodium monofluorophosphate dentifrices; although few details of this study have been reported, they observed numerically greater caries benefits with the sodium fluoride products. The most recent report comparing these fluoride systems was by Koch and coworkers.[157] These investigators compared the same two fluorides at the conventional concentration and included an additional sodium fluoride dentifrice which contained only one-fourth the usual amount of fluoride. At termination, no significant differences were found between any of the products although slightly lower caries increments were observed with the sodium monofluorophosphate composition. Thus, four of the five studies comparing these two fluoride systems found the sodium fluoride to be more effective and in two of the studies the differences were significant. It should also be noted that all instances in which no significant difference was observed between the sodium fluoride and sodium monofluorophosphate dentifrices, the subjects were simultaneously receiving sodium fluoride rinses.[119,156,157]

Table 9–28.   Clinical Comparison of SnF_2-$Ca_2P_2O_7$ and NaF-SiO_2 Dentifrices

Study Reported	More Effective	Significant Difference
Zacherl (1981)[65]	NaF	Yes
Beiswanger et al. (1981)[154]	NaF	Yes

Table 9–29.   Clinical Dentifrice Studies Comparing Na_2PO_3F and Compatible NaF Formulations

Study Reported	Abrasive System		Significant Difference
	NaF	Na_2PO_3F	
Gerdin (1972)[87]	Acrylic	$CaCO_3$	Yes
Forsman (1974)[119]	Silica	Silica	No
Edlund & Koch (1977)[155]	Silica	$CaHPO_4$/$CaCO_3$	Yes
Edward & Torell (1978)[156]	Silica	$CaCO_3$	No
	Silica	Silica	No
Koch et al. (1982)[157]	Silica	$CaHPO_4$	No

During the three decades involved in the conduct of these numerous clinical trials, scientists continued to investigate fundamental factors related to both the etiology of caries and the mechanism of action of fluoride. It is important that we recognize that our understanding of these processes has markedly changed during that time, since our current knowledge of these processes is helpful in explaining the clinical results observed with fluoride dentifrices.

Instead of the caries process being initiated by simple demineralization of the enamel surface, it is now generally accepted that this process (Fig. 9–1) begins with a subsurface demineralization of enamel just beneath a relatively intact or well-mineralized enamel surface.[158,159] Furthermore, various types of information suggest that this same etiological process is involved in all types of coronal lesions—smooth surface as well as pits and fissures—in both the primary and permanent dentition. It is apparent from these changes in our understanding of the caries process that fluoride dentifrices should be recommended for adults as well as children. In addition, it is now well accepted that these subsurface incipient lesions may be remineralized by saliva as a normal reparative process (Fig. 9–2), and, of particular importance, the rate and amount of remineralization are markedly increased by the presence of fluoride[160] (Fig. 9–3).

While many questions remain to be answered regarding the mechanism of action of fluoride, particularly relating to more complex fluorides such as monofluorophosphate, it is apparent that the original concepts of the '50s and '60s are in question. In particular, the suggestion that fluoride reacted with sound, fully maturated enamel to make the surface more resistant to acid attack is no longer accepted as the primary mechanism. The fact that fluoride accumulates in plaque is well-known but its subsequent fate remains unclear; while it is now known not to be present in bacteriocidal concentrations, it has been shown to be reactive with enamel when the pH of the plaque falls below the critical value. Current knowledge indicates the primary mechanism of action of fluoride is related to its diffusion into the

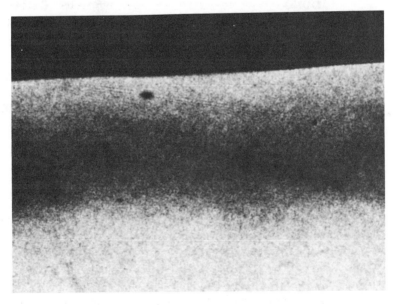

Fig. 9–1. Photograph of an incipient, subsurface lesion.

AN EQUILIBRIUM SITUATION

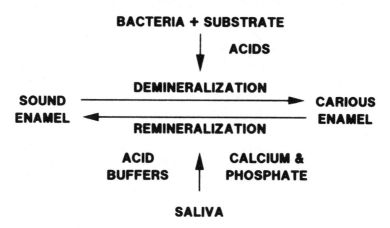

Fig. 9–2. Remineralization-Demineralization Equilibrium.

THE CARIES PROCESS—
AN EQUILIBRIUM SITUATION

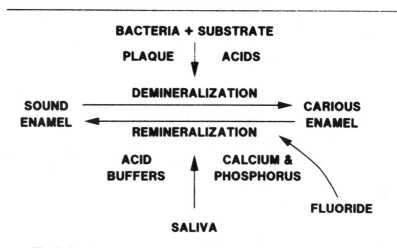

Fig. 9–3. Remineralization-Demineralization Equilibrium with Fluoride.

early subsurface lesions and its ability to enhance the remineralization of these areas.[160–162] Various reports during the past few years suggest that the cariostatic ability of fluoride provided by fluoride-containing dentifrices similarly involves the deposition of fluoride and remineralization of incipient enamel lesions.

A report by Cilley and coworkers[163] involved fluoride analyses of exfoliated deciduous teeth from children participating in a clinical caries trial of fluoride dentifrices (Table 9–30). The results indicated that there was no significant increase in the fluoride content of sound enamel surfaces. In contrast, the analyses of incipient carious lesions indicated a significant increase in fluoride content with both fluoride

Table 9–30. Fluoride Uptake in Enamel of Children Using NaF and SnF$_2$ Dentifrices[163]

Dentifrice Used	Fluoride Content (ppm)	
	Sound Enamel	Incipient Lesions
Placebo	364	731
SnF$_2$-Ca$_2$P$_2$O$_7$	373	878
NaF-SiO$_2$	379	1148

dentifrices and a greater deposition of fluoride with the sodium fluoride formulation. It is worthy to note that these results paralleled the caries findings in the clinical caries study.[65]

An in vivo investigation by Mobley[164] involving the deposition of fluoride in removable enamel chips containing artificial incipient lesions and contained in appliances worn by patients showed similar results (Table 9–31). Two in vivo studies conducted in our laboratories[165] resulted in similar observations and again significantly greater deposition was observed with a sodium fluoride formulation as compared to the stannous fluoride composition (Table 9–32).

Similar investigations reported by Mellberg and coworkers[166,167] have also indicated that a variety of sodium monofluorophosphate dentifrice formulations produce a significant increase in the fluoride content of incipient lesions (Table 9–33). They further suggest that the presence of calcium enhanced this deposition of fluoride although we have been unable to confirm this in our in vivo studies. With regard to the comparative deposition of fluoride[166,168] from different fluoride salts, Mellberg and Chomicki[166] observed numerically greater fluoride deposition with a sodium fluoride-silica formulation than with a sodium monofluorophosphate-dicalcium phosphate dentifrice but the differences were not statistically significant. In a study which we conducted using essentially the same dentifrices,[165] however, significantly

Table 9–31. In Vivo Fluoride Uptake in Artificial Incipient Lesions From NaF and SnF$_2$ Dentifrices[164]

Dentifrice Studied	Fluoride Content of Lesions (μg/cm^2)
Placebo	4.6
SnF$_2$-Ca$_2$P$_2$O$_7$	8.4
NaF-SiO$_2$	16.0

Table 9–32. In Vivo Fluoride Uptake in Artificial Incipient Lesions from NaF and SnF$_2$ Dentifrices[165]

Dentifrice Studied	Content of Lesions (μg F/cm^2)	
	Study #1	Study #2
Placebo	4.1	4.7
SnF$_2$-Ca$_2$P$_2$O$_7$	6.4	5.9
NaF-SiO$_2$	8.6	9.0

greater amounts of fluoride deposition were observed with the sodium fluoride formulation after usage periods of both 4 and 9 weeks (Table 9–34).

A similar in vivo study was conducted in our laboratories in which the panelists used each of three different dentifrices for 4 weeks (Table 9–35). The fluoride dentifrices differed in total fluoride content as well as the source of fluoride. The formulation containing both fluoride compounds was essentially the same as that evaluated clinically by Hodge and coworkers.[111] Again, both fluoride dentifrices resulted in signficant increases in the fluoride content of the lesions and significantly greater amounts of fluoride deposition in incipient lesions were observed with the conventional sodium fluoride dentifrice. Collectively, these studies indicate that the use of sodium fluoride dentifrices results in greater fluoride deposition in incipient lesions than is obtained with stannous fluoride or sodium monofluorophosphate systems.

Several recent reports have also assessed the ability of fluoride dentifrices to remineralize incipient lesions both in vitro and in vivo using a variety of parameters including microradiography, polarized light microscopy, and changes in the hardness of the lesions.[162,169–173] Most of the pertinent studies have involved sodium monofluorophosphate and sodium fluoride dentifrices. For example, Mellberg[173] noted in an in vivo study that brushing 3 times daily for 2 months with a sodium monofluorophosphate dentifrice, as compared to a placebo product, resulted in an 11% increase in the mineral content of incipient lesions. Similarly, in an in vivo study we have observed that 2 weeks of twice daily toothbrushing with a sodium fluoride dentifrice

Table 9–33. In Vivo Fluoride Uptake in Artificial Incipient Lesions from NaF and Na_2PO_3F Dentifrices (Mellberg & Chomicki[167])

Dentifrice Studied	Fluoride Content of Lesions (ppm)
Placebo	~1700
Na_2PO_3F-$CaHPO_4$	~3800
NaF-SiO_2	~4600

Table 9–34. In Vivo Fluoride Uptake in Artificial Incipient Lesions With NaF and Na_2PO_3F Dentifrices

Dentifrice Studied	Content of Lesions (μg F/cm^2)	
	4 Weeks	9 Weeks
Na_2PO_3F-$CaHPO_4$	8.8	9.0
NaF-SiO_2	12.8	14.3

Table 9–35. In Vivo Fluoride Uptake with NaF and Na_2PO_3F/NaF Dentifrices

Dentifrice Studied	Fluoride System	Fluoride Content of Lesions ($\mu g/cm^2$)
Placebo	None	2.28
Na_2PO_3F/NaF	Na_2PO_3F (0.10% F) NaF (0.045% F)	3.47
NaF	NaF (0.11% F)	5.11

Table 9–36. In Vivo Fluoride Uptake and Remineralization in Incipient Lesions with NaF Dentifrice

Dentifrice Provided	Fluoride Uptake ($\mu g/cm^2$)	Increased Hardness (VHN)
Placebo	2.2	4.1
NaF-SiO$_2$	6.0	20.4

results in a pronounced increase in the hardness of incipient enamel lesions as compared to that observed when the patients similarly used a placebo formulation (Table 9–36). Furthermore, the increased hardness paralleled an increase in fluoride content of the lesions and an increase in the hardness of incipient lesions, which has been shown to reflect remineralization.[174] From these investigations, it is clearly apparent that both of these fluoride compounds facilitate the remineralization of incipient lesions.

With regard to the comparative ability of these two compounds to facilitate this process, in vitro studies generally indicate a greater efficiency of sodium fluoride[171,172]; however, it must be recognized that in vitro data must be interpreted with some caution, particularly with regard to complex fluorides where the mechanism of action is not clearly understood.

On the other hand, the results of a clinical investigation by Gerdin and Serneke[175] are pertinent and corroborate the in vitro investigations. Gerdin and Serneke provided children with either sodium fluoride dentifrices having sodium bicarbonate or acrylic particles as the abrasive or with a sodium monofluorophosphate-calcium carbonate dentifrice. Weekly supervised brushing plus home use of the dentifrices was performed for 15 months. Remineralization was determined on the basis of changes in enamel continuity of an initially placed enamel scratch as evidenced by an SEM replica procedure. Although the investigators considered remineralization as a surface phenomenon, the results indicated significantly greater remineralization occurred with the neutral sodium fluoride-acrylic particle formulation.

The results of these studies of fluoride deposition in incipient caries as well as the role of fluoride in the remineralization of these lesions provide a reasonable basis for explaining the cariostatic activity of fluoride-containing dentifrices. These data also help to explain the apparent differences in the clinical efficacy of different fluoride systems. It is also apparent from the clinical data cited earlier that the mere presence of soluble or available fluoride in a dentifrice is not enough to explain the level of cariostatic activity of the system but instead the bioavailability, or the biological utilization, of that fluoride as evidenced by fluoride deposition and remineralization of incipient lesions, is necessary to achieve cariostatic activity.

CONCLUDING REMARKS

It is now time to address the original question "Are all fluoride dentifrices the same?" Having reviewed the available information it should be clearly apparent that the answer is "No, all fluoride dentifrices are not the same." It is quite obvious from this review of the literature that significant cariostatic benefits are provided by a number of fluoride dentifrices, namely, stannous fluoride used with calcium pyrophosphate, insoluble sodium metaphosphate or silica; amine fluoride used with

Table 9–37. Clinically Effective Fluoride-Abrasive Systems Used in Dentifrices

Fluoride	Abrasives
SnF_2	$Ca_2P_2O_7$, $(NaPO_3)_x$ $\pm$ $CaHPO_4$, SiO_2
Amine F	$(NaPO_3)_x$
Na_2PO_3F	$CaHPO_4$, $(NaPO_3)_x$, Al_2O_3, $CaCO_3$, SiO_2
NaF	Modified $Ca_2P_2O_7$, $(NaPO_3)_x$, Acrylic, SiO_2

insoluble sodium metaphosphate; sodium monofluorophosphate used with a wide variety of abrasive systems; and sodium fluoride used with a special calcium pyrophosphate, insoluble sodium metaphosphate, acrylic particles or silica (Table 9–37).

The degree of effectiveness, however, appears to vary for different formulations. Since the major systems of interest in the United States today are formulations containing either sodium monofluorophosphate or sodium fluoride, it is reasonable to ask if these particular systems are different. In spite of the notable absence of head-to-head comparative clinical trials of all possible dentifrice formulations, the available body of data convincingly indicate a difference in efficacy. Since this is a critical point, let us quickly review the data that justify this conclusion:

(1) Clinical trials have demonstrated that sodium monofluorophosphate compositions and stannous fluoride dentifrices are comparable in effectiveness;[32,49,51,55,58,71,76,77]

(2) Two clinical studies[65,154] have shown that a sodium fluoride-silica dentifrice is significantly more effective than the stannous fluoride-calcium pyrophosphate product used as point of reference for many years;

(3) The results of one clinical study[58] demonstrated that a sodium fluoride-improved calcium pyrophosphate dentifrice with an intermediate level of available fluoride was similar to a sodium monofluorophosphate-calcium pyrophosphate formulation;

(4) In comparing various sodium monofluorophosphate dentifrices with sodium fluoride dentifrices formulated with highly compatible abrasives, namely, silica or acrylic particles, four out of five clinical trials demonstrated numerically greater effectiveness of sodium fluoride, and in two of the studies the differences were statistically significant;[87,119,155–157]

(5) A clinical study[118] in which one-half of the sodium monofluorophosphate was replaced with sodium fluoride resulted in numerically greater cariostatic activity than when all the fluoride was provided by sodium monofluorophosphate;

(6) Explanatory data from independent studies regarding the deposition of fluoride in incipient carious lesions indicate that a greater amount of fluoride deposition was obtained with a sodium fluoride dentifrice than was obtained with conventional sodium monofluorophosphate dentifrices;

(7) Explanatory data from in vitro and in vivo studies of the remineralization of incipient lesions and enamel irregularities indicate a greater efficiency of highly compatible neutral sodium fluoride formulations than was apparent with sodium monofluorophosphate dentifrices.

On the basis of this collective body of data it is concluded that all fluoride dentifrices are not equal in terms of the magnitude of their cariostatic benefits and that at the present time sodium fluoride dentifrices, formulated so as to provide a high level of available and biologically active fluoride, are the most effective.

REFERENCES

1. Bibby, B.G.: A test of the effect of fluoride-containing dentifrices on dental caries. J. Dent. Res., *24*:297–303, 1945.
2. Bibby, B.G., and Wellock, W.D.: Unpublished report, 1948. *Cited in* Fluoridation as a Public Health Measure. Edited by J.H. Shaw. AAAS, 1954, pp. 158–159.
3. Wellock, W.D., and Bibby, B.G.: Unpublished report, 1948. *Cited in* Fluoridation as a Public Health Measure. Edited by J.H. Shaw. AAAS, 1954, pp. 158–159.
4. Winkler, K.C., Backer-Dirks, O., and van Amerongen, J.: A reproducible method for caries evaluation. Test in a therapeutic experiment with a fluorinated dentifrice. Br. Dent. J., *95*:119–124, 1953.
5. Muhler, J.C., Radike, A.W., Nebergall, W.H., and Day, H.G.: A comparison between the anticariogenic effects of dentifrices containing stannous fluoride and sodium fluoride. J. Am. Dent. Assoc., *51*:556–559, 1955.
6. Muhler, J.C.: Effect on dental caries of a dentifrice containing stannous fluoride and dicalcium phosphate. J. Dent. Res., *36*:399–402, 1957.
7. Kyes, F.M., Overton, N.J., and McKean, T.W.: Clinical trials of caries inhibitory dentifrices. J. Am. Dent. Assoc., *63*:189–193, 1961.
8. Brudevold, F., and Chilton, N.W.: Comparative study of a fluoride dentifrice containing soluble phosphate and a calcium-free abrasive: Second-year report. J. Am. Dent. Assoc., *72*:889–894, 1966.
9. Grøn, P., and Brudevold, F.: The effectiveness of NaF dentifrices. J. Dent. Child., *34*:122–127, 1967.
10. Manly, R.S.: A structureless recurrent deposit on teeth. J. Dent. Res., *22*:479–486, 1943.
11. Vallotton, C.F.: An acquired pigmented pellicle of the enamel surface. II. Clinical and histologic studies. J. Dent. Res., *24*:171–181, 1945.
12. McCauley, H.B., et al.: Clinical efficacy of powder and paste dentifrices. J. Am. Dent. Assoc., *33*:993–997, 1946.
13. Kitchin, P.C., and Robinson, H.B.G.: How abrasive need a dentifrice be? J. Dent. Res., *27*:501–506, 1948.
14. Dudding, N.J., Dahl, L.O., and Muhler, J.C.: Patient reactions to brushing teeth with water, dentifrice or salt and soda. J. Periodontol., *31*:386–392, 1960.
15. Lobene, R.R.: Effect of dentifrices on tooth stains with controlled brushing. J. Am. Dent. Assoc., *77*:849–855, 1968.
16. Lobene, R.R.: Clinical studies of the cleaning functions of dentifrices. J. Am. Dent. Assoc., *105*:798–802, 1982.
17. Hefferren, J.J.: A laboratory method for assessment of dentifrice abrasivity. J. Dent. Res., *55*:563–573, 1976.
18. Volpe, A.R., et al.: A long-term clinical study evaluating the effect of two dentifrices on oral tissues. J. Periodontol., *46*:113–118, 1975.
19. Schiff, T., and Volpe, A.R.: A two-year clinical study comparing the effect of dentifrices on selected dental materials. J. Oral Rehabil., *2*:407–412, 1975.
20. Bergstrom, J., and Lavstedt, S.: An epidemiologic approach to toothbrushing and dentin abrasion. Community Dent. Oral Epidemiol., *7*:57–64, 1979.
21. Ericsson, Y.: Fluorides in dentifrices. Investigations using radioactive fluorine. Acta Odontol. Scand., *19*:41–77, 1961.
22. Muhler, J.C., Radike, A.W., Nebergall, W.H., and Day, H.G.: The effect of a stannous fluoride-containing dentifrice on caries reduction in children. J. Dent. Res., *33*:606–612, 1954.
23. Muhler, J.C., Radike, A.W., Nebergall, W.H., and Day, H.G.: Effect of a stannous fluoride-containing dentifrice on caries reduction in children. II. Caries experience after one year. J. Am. Dent. Assoc., *50*:163–166, 1955.
24. Muhler, J.C., and Radike, A.W.: Effect of a dentifrice containing stannous fluoride on dental caries in adults. II. Results at the end of two years of unsupervised use. J. Am. Dent. Assoc., *55*:196–198, 1957.
25. Muhler, J.C.: The effect of a modified stannous fluoride-calcium pyrophosphate dentifrice on dental caries in children. J. Dent. Res., *37*:448–450, 1958.
26. Jordan, W.A., and Peterson, J.K.: Caries-inhibiting value of a dentifrice containing stannous fluoride. Final report of a two-year study. J. Am. Dent. Assoc., *58*:42–46, 1959.
27. Hill, T.J.: Fluoride dentifrices. J. Am. Dent. Assoc., *59*:1121–1127, 1959.
28. Peffley, G.E., and Muhler, J.C.: The effect of a commercially available stannous fluoride denti-

frice under controlled brushing habits on dental caries incidence in children: Preliminary report. J. Dent. Res., *39*:871–875, 1960.

29. Muhler, J.C.: Combined anticariogenic effect of a single stannous fluoride solution and the unsupervised use of a stannous fluoride-containing dentifrice. II. Results at the end of two years. J. Dent. Res., *39*:955–958, 1960.

30. Muhler, J.C.: Effect of a stannous fluoride dentifrice on caries reduction in children during a three-year study period. J. Am. Dent. Assoc., *64*:216–224, 1962.

31. Bixler, D., and Muhler, J.C.: Experimental clinical human caries test design and interpretation. J. Am. Dent. Assoc., *65*:482–490, 1962.

32. Finn, S.B., and Jamison, H.C.: A comparative clinical study of three dentifrices. J. Dent. Child., *30*:17–25, 1963.

33. Henriques, B.L., Frankl, S.N., and Alman, J.E.: *Cited In* Evaluation of Cue tooth paste. J. Am. Dent. Assoc., *69*:197–198, 1964. (Also in Volpe, A.R.: Dentifrices and Mouthrinses. Chap. 10. *In* A Textbook of Preventive Dentistry. Edited by R.E. Stallard. 2nd Ed. Philadelphia, W.B. Saunders Company, 1982.)

34. Mergele, M., Jennings, R.E., and Gasser, E.B.: *Cited In* Evaluation of Cue tooth paste. J. Am. Dent. Assoc., *69*:197–198, 1964. (Also in Volpe, A.R.: Dentifrices and Mouthrinses. Chap. 10. *In* A Textbook of Preventive Dentistry. Edited by R.E. Stallard. 2nd Ed. Philadelphia, W.B. Saunders Company, 1982.)

35. Torell, P., and Ericsson, Y.: Two-year clinical tests with different methods of local caries-preventive flourine application in Swedish school-children. Acta Odontol. Scand., *23*:287–312, 1965.

36. Gish, C.W., and Muhler, J.C.: Effectiveness of a SnF_2-$Ca_2P_2O_7$ dentifrice on dental caries in children whose teeth calcified in a natural fluoride area. II. Results at the end of 24 months. J. Am. Dent. Assoc., *73*:853–855, 1966.

37. Horowitz, H.S., Law, F.E., Thompson, M.B., and Chamberlin, S.R.: Evaluation of a stannous fluoride dentifrice for use in Dental Public Health Programs. I. Basic findings. J. Am. Dent. Assoc., *72*:408–422, 1966.

38. Horowitz, H.S., and Thompson, M.B.: Evaluation of a stannous fluoride dentifrice for use in Dental Public Health Programs. III. Supplementary findings. J. Am. Dent. Assoc., *74*:979–986, 1967.

39. Halikis, S.E.: A pilot study on the effectiveness of a stannous fluoride dentifrice on dental caries in children. Aust. Dent. J., *11*:336–337, 1966.

40. Thomas, A.E., and Jamison, H.C.: Effect of SnF_2 dentifrices on caries in children: Two-year clinical study of supervised brushing in children's homes. J. Am. Dent. Assoc., *73*:844–852, 1966.

41. Lehnhoff, R.W., et al.: Clinical measurement of the effect of an anticaries dentifrice by three examiners. Int. Assn. Dent. Res., Preprinted Abs. #246, 1966.

42. Bixler, D., and Muhler, J.C.: Effectiveness of a stannous fluoride-containing dentifrice in reducing dental caries in a boarding school environment. J. Am. Dent. Assoc., *72*:653–658, 1966.

43. Jackson, D., and Sutcliffe, P.: Clinical testing of a stannous fluoride-calcium pyrophosphate dentifrice in Yorkshire school children. Br. Dent. J., *123*:40–48, 1967.

44. James, P.M.C., and Anderson, R.J.: Clinical testing of a stannous fluoride-calcium pyrophosphate dentifrice in Buckinghamshire school children. Br. Dent. J., *123*:33–39, 1967.

45. Slack, G.L., Berman, D.S., Martin, W.J., and Hardie, J.M.: Clinical testing of a stannous fluoride-calcium pyrophosphate dentifrice in Essex school girls. Br. Dent. J., *123*:26–33, 1967.

46. Weisenstein, P.R., and Lehnhoff, R.W.: A clinical evaluation of an anticaries dentifrice comparing conventional and radiographic measurements. Int. Assn. Dent. Res., Preprinted Abs. #425, 1967.

47. Onishi, E.: Effect of stannous fluoride dentifrice on caries reduction of school children. Jpn. J. Dent. Health, *17*:68–74, 1967 (Oral Res. Abs. *3*:497, 1968).

48. Peterson, J.K., and Williamson, L.: Three-year caries inhibition of a sodium fluoride acid orthophosphate dentifrice compared with a stannous fluoride dentifrice and a nonfluoride dentifrice. Int. Assn. Dent. Res., Preprinted Abs. #255:101, 1968.

49. Mergele, M.: Report II. An unsupervised brushing study on subjects residing in a community with fluoride in the water. Bull. Acad. Med. NJ, *14*:251–255, 1968.

50. Zacherl, W.A.: A clinical evaluation of sodium fluoride and stannous fluoride dentifrices. Int. Assn. Dent. Res., Preprinted Abs. #253:101, 1968.

51. Frankl, S.N., and Alman, J.E.: Report of a three-year clinical trial comparing a toothpaste containing sodium monofluorophosphate with two marketed products. J. Oral Ther. & Pharm., *4*:443–450, 1968.

52. Zacherl, W.A.: Clinical evaluation of a sarcosinate dentifrice. Int. Assn. Dent. Res., Preprinted Abs. #339:133, 1970.

53. Zacherl, W.A., and McPhail, C.W.B.: Final report on the efficacy of a stannous fluoride-calcium pyrophosphate dentifrice. J. Can. Dent. Assoc., *36*:262–264, 1970.

54. Muhler, J.C.: A clinical comparison of fluoride and antienzyme dentifrices. J. Dent. Child., *37*:501–502, 511–514, 1970.

55. Onisi, M., and Tani, H.: Clinical test on the caries-preventive effect of two kinds of fluoride dentifrices. Jpn. J. Dent. Health, *20*:105–111, 1970 (Oral Res. Abs. *7*:399, 1972).

56. Slack, G.L., Bulman, J.S., and Osborn, J.F.: Clinical testing of fluoride and nonfluoride containing dentifrices in Hounslow school children. Br. Dent. J., *130*:154–158, 1971.

57. Gish, C.W., and Muhler, J.C.: Effectiveness of a stannous fluoride dentifrice on dental caries. J. Dent. Child., *38*:211–214, 1971.

58. Zacherl, W.A.: Clinical evaluation of neutral sodium fluoride, stannous fluoride, sodium monofluorophosphate and acidulated fluoride–phosphate dentifrices. J. Can. Dent. Assoc., *38*:35–38, 1972.

59. Zacherl, W.A.: Clinical evaluation of an aged stannous fluoride-calcium pyrophosphate dentifrice. J. Can. Dent. Assoc., *38*:155–157, 1972.

60. Zacherl, W.A.: A clinical evaluation of a stannous fluoride and a sarcosinate dentifrice. J. Dent. Child., *40*:451–453, 1973.

61. Ringelberg, M.L., Webster, D.B., Dixon, D.O., and LaZotte, D.C.: The caries preventive effect of amine fluorides and inorganic fluorides in a mouthrinse or dentifrice after 30 months of use. J. Am. Dent. Assoc., *98*:202–208, 1979.

62. Fogels, H.R., Alman, J.E., Meade, J.J., and O'Donnell, J.P.: The relative caries-inhibiting effects of a stannous fluoride dentifrice in a silica gel base. J. Am. Dent. Assoc., *99*:456–459, 1979.

63. Lu, K.H., Hanna, J.D., and Peterson, J.K.: Effect on dental caries of a stannous fluoride-calcium pyrophosphate dentifrice in an adult population: One-year results. Pharmacol. Ther. Dent., *5*:11–16, 1980.

64. Abrams, R.G., and Chambers, D.W.: Caries-inhibiting effect of a stannous fluoride silica gel dentifrice: A three-year clinical study. Clin. Prev. Dent., *2*:22–27, 1980.

65. Zacherl, W.A.: A three-year clinical caries evaluation of the effect of a sodium fluoride-silica abrasive dentifrice. Pharmacol. Ther. Dent., *6*:1–7, 1981.

66. American Dental Association, Council on Dental Therapeutics: Evaluation of Crest toothpaste. J. Am. Dent. Assoc., *61*:272–274, 1960.

67. American Dental Association, Council on Dental Therapeutics: Reclassification of Crest toothpaste. J. Am. Dent. Assoc., *69*:195–196, 1964.

68. Slack, G.L., and Martin, W.J.: The use of a dentifrice containing stannous fluoride in the control of dental caries. Br. Dent. J., *117*:275–280, 1964.

69. Fullmer, J., Volpe, A.R., Apperson, L.D., and Kiraly, J.: Unpublished data. *Cited In* Volpe, A.R.: Dentifrices and Mouthrinses. Chap. 10. *In* A Textbook of Preventive Dentistry. Edited by R.E. Stallard. 2nd Ed. Philadelphia, W.B. Saunders Company, 1982.

70. Segal, A.H., Stiff, R. H., George, W.A., and Picozzi, A.: Cariostatic effect of a stannous fluoride-containing dentifrice on children: two-year report of a supervised toothbrushing study. J. Oral Ther. & Pharm., *4*:175–180, 1967.

71. Naylor, M.N., and Emslie, R.D.: Clinical testing of stannous fluoride and sodium monofluorophosphate dentifrices in London school children. Br. Dent. J., *123*:17–23, 1967.

72. Ashley, F.P., Naylor, M.M., and Emslie, R.D.: Stannous fluoride and sodium monofluorophosphate dentifrices. Clinical testing in London school children—radiological findings. Br. Dent. J., *127*:125–128, 1969.

73. American Dental Association, Council on Dental Therapeutics: Evaluation of Cue toothpaste. J. Am. Dent. Assoc., *69*:197–198, 1964.

74. American Dental Association, Council on Dental Therapeutics: Evaluation of Super Stripe toothpaste. J. Am. Dent. Assoc., *72*:1515, 1966.

75. Slack, G.L., Berman, D.S., Martin, W.J., and Young, J.: Clinical testing of a stannous fluoride-insoluble metaphosphate dentifrice in Kent school girls. Br. Dent. J., *123*:9–16, 1967.

76. Fanning, E.A., Gotjamanos, T., and Vowles, N.J.: The use of fluoride dentifrices in the control of dental caries: Methodology and results of a clinical trial. Aust. Dent. J., *13*:201–206, 1968.

77. Mergele, M.: Report I. A supervised brushing study in State Institution schools. Bull. Acad. Med. NJ, *14*:247–250, 1968.

78. American Dental Association, Council on Dental Therapeutics: Evaluation of Fact toothpaste. J. Am. Dent. Assoc., *71*:930–931, 1965.

79. Homan, B.T., and Messer, H.H.: The comparative effect of three fluoride dentifrices on clinical dental caries in Brisbane schoolchildren. Preliminary report. J. Dent. Res., *48*:1094, 1969.

80. American Dental Association, Council on Dental Therapeutics. Council accepts Aim. J. Am. Dent. Assoc., *99*:699, 1979.

81. Howell, C.L., and Muhler, J.C.: Effect of topically applied stannous chlorofluoride on the dental caries experience in children. II. Results two years after initial treatment. J. Am. Dent. Assoc., *55*:493–495, 1957.

82. Gish, C.W., Muhler, J.C., and Howell, C.L.: The effect of topically applied potassium fluorostannite on the dental caries experience in children. III. Results at the end of three years. J. Dent. Res., *38*:881–882, 1959.

83. Muhler, J.C., Bixler, D., and Stookey, G.K.: The clinical effectiveness of stannous hexafluoro-zirconate as an anticariogenic agent. J. Am. Dent. Assoc., 76:558–563, 1968.

84. Held, A.J., and Spirgi, M.: Clinical experimentation with fluoridated dentifrice. Schweiz. Mschr. Zahnheilk., 75:883–902, 1965. (Oral Res. Abs. 1:409, 1966.)

85. Held, A.J., and Spirgi, M.: Three years of clinical observations with fluoridated dentifrices. Bull. Group Int. Rech. Sci. Stomatol. Odontol., 11:539–570, 1968 (Oral Res. Abs. 4:1085, 1964).

86. Geiger, L., Kunzel, W., and Treide, A.: Comparative clinical-radiological examination of caries decrease after supervised oral hygiene with ammonium fluoride. Dtsch. Stomatol., 21:135–139, 1971.

87. Gerdin, P.O.: Studies in Dentifrices, VI: The inhibitory effect of some grinding and nongrinding fluoride dentifrices on dental caries. Swed. Dent. J., 65:521–532, 1972.

88. Koch, G.: Comparison and estimation of effect on caries of daily supervised toothbrushing with a dentifrice containing sodium fluoride and a dentifrice containing potassium fluoride and manganese chloride. A three-year clinical test. Odont. Revy, 23:341–354, 1972.

89. Gerdin, P.O.: Studies in dentifrices, VIII: Clinical testing of an acidulated, nongrinding dentifrice with reduced fluorine contents. Swed. Dent. J., 67:283–297, 1974.

90. Marthaler, T.M.: Caries-inhibition after seven years of unsupervised use of an amine fluoride dentifrice. Br. Dent. J., 124:510–515, 1968.

91. von Patz, J., and Naujoks, R.: The prophylactic anticaries effect of an amine fluoride-containing dentifrice on adolescents after unsupervised use for 3 years. Dtsch. Zahnaerztl. Z., 25:617–625, 1970. (Oral Res. Abs. 6:204, 1971.)

92. Marthaler, T.M.: Caries-inhibition by an amine fluoride dentifrice. Results after 6 years in children with low caries activity. Helv. Odontol. Acta, 18 (Suppl. VIII):35–44, 1974.

93. Kunzel, W., Franke, W., and Treide, A.: Klinisch-Rontgenologische Paralleluberwachung einer Langsschnittstudie zum Nachweis der Karieshemmenden Effektivitat 7 Jahre Lokal angewandeten Aminfluorids im Doppelblindtest. Zahn Mund. u. Keiferheilk, 65:626–637, 1977.

94. Cahen, P.M., Frank, R.M., Turlot, J.C., and Jung, M.T.: Comparative unsupervised clinical trial on caries inhibition effect of monofluorophosphate and amine fluoride dentifrices after 3 years in Strasbourg, France. Community Dent. Oral Epidemiol., 10:238–241, 1982.

95. Kinkel, H.J., and Stolte, G.: On the effect of a sodium monofluorophosphate and bromochloro-phene-containing toothpaste in a chronic animal experiment and on caries in children during a two-year period of unsupervised use. Dtsch. Zahnaerztl. Z., 22:455–460, 1968.

96. Patz, J., and Naujoks, R.: Clinical investigation of a fluoride-containing dentifrice in adults. Results of a two-year unsupervised study. Dtsch. Zahnaerztl. Z., 7:614–621, 1969.

97. Barlage, B., Buhe, H., and Buttner, W.: A 3-year clinical dentifrice trial using different fluoride levels: 0.8 and 1.2% sodium monofluorophosphate. Caries Res., 15:185 (Abs. #18), 1981.

98. Møller, I.J., Holst, J.J., and Sorensen, E.: Caries-reducing effect of a sodium monofluorophosphate dentifrice. Br. Dent. J., 124:209–213, 1968.

99. Thomas, A.E., and Jamison, H.C.: Effect of a combination of two cariostatic agents on caries in children: Two-year clinical study of supervised brushing in children's homes. J. Am. Dent. Assoc., 81:118–124, 1970.

100. Peterson, J.K.: A supervised brushing trial of sodium monofluorophosphate dentifrices in a fluoridated area. Caries Res., 13:68–72, 1979.

101. Glass, R.L., Peterson, J.K., and Bixler, D.: The effects of changing caries prevalence and diagnostic criteria on clinical caries trials. Caries Res., 17:145–151, 1983.

102. Blinkhorn, A.S., Holloway, P.J., and Davies, T.G.H.: Combined effects of a fluoride dentifrice and mouthrinse on the incidence of dental caries. Community Dent. Oral Epidemiol., 11:7–11, 1983.

103. Ashley, F.P., Mainwaring, P.J., Emslie, R.D., and Naylor, M.N.: Clinical testing of a mouthrinse and a dentifrice containing fluoride. A two-year supervised study in school children. Br. Dent. J., 143:333–338, 1977.

104. Takeuchi, M., Shimizu, T., Kawasaki, T., and Kizu, T.: Caries prevention by a sodium monofluorophosphate dentifrice. Jpn. J. Dent. Health, 18:26–38, 1968. (Oral Res. Abs. 4:489, 1969).

105. Kinkel, H.J., and Raich, R.: Caries inhibition after 5 years application of a dentifrice containing Na$_2$PO$_3$F. Schweiz. Monatsschr. Zahnheilk., 84:1245–1247, 1974.

106. Kinkel, H.J., Stolte, G., and Weststrate, J.: Study of the effect of a toothpaste containing fluorophosphate on the dentition of children. Schweiz. Monatsschr. Zahnheilk., 84:577–589, 1974.

107. Niwa, T., Baba, K., and Niwa, N.: The caries-preventive effects of dentifrices containing sodium monofluorophosphate, sodium monofluorophosphate plus dextranase, and sodium monofluorophosphate plus sodium phosphate. Jpn. J. Dent. Health, 25:30–52, 1975.

108. Rijnbeek, P.L.C.A., and Weststrate, J.: Development and study of the efficacy of a fluoridated dentifrice. Nederl. T. Tandheelk, 83:123–128, 1976 (Oral Res. Abs., 12:498, 1977).

109. Kinkel, H.J., Raich, R., and Muller, M.: Die Karieshemmung einer Na$_2$PO$_3$F-Zahnpasta nach 7 Jahren Applikation. Schweiz. Mschr. Zahnheilk., 87:1218–1220, 1977.

110. Andlaw, R.J., and Tucker, G.J.: A three-year clinical trial of a dentifrice containing 0.8 percent sodium monofluorophosphate in an aluminum oxide trihydrate base. Br. Dent. J., *138*:426–432, 1975.

111. Hodge, H.C., Holloway, P.J., Davies, T.G.H., and Worthington, H.V.: Caries prevention by dentifrices containing a combination of sodium monofluorophosphate and sodium fluoride. Report of a 3-year clinical trial. Br. Dent. J., *149*:201–204, 1980.

112. Murray, J.J., and Shaw, L.: A 3-year clinical trial into the effect of fluoride content and toothpaste abrasivity on the caries inhibitory properties of a dentifrice. Community Dent. Oral Epidemiol., *8*:46–51, 1980.

113. Andlaw, R.J., Palmer, J.D., King, J., and Kneebone, S.B.: Caries-preventive effects of toothpastes containing monofluorophosphate and trimetaphosphate: A 3-year clinical trial. Community Dent. Oral Epidemiol., *11*:143–147, 1983.

114. Torell, P.: The use of fluoride toothpaste combined with fluoride rinsing every two weeks. Sver. Tandlak. Forb. Tidn., *61*:873–875, 1969.

115. Peterson, J.K., Williamson, L.D., and Casad, R.: Caries inhibition with MFP-calcium carbonate dentifrice in fluoridated area. Int. Assn. Dent. Res., Preprinted Abs. #L338, 1975.

116. Mainwaring, P.J., and Naylor, M.N.: A three-year clinical study to determine the separate and combined caries-inhibiting effects of sodium monofluorophosphate toothpaste and an acidulated phosphate-fluoride gel. Caries Res., *12*:202–212, 1978.

117. Glass, R.L., and Shiere, F.R.: A clinical trial of a calcium carbonate base dentifrice containing 0.76% sodium monofluorophosphate. Caries Res., *12*:284–289, 1978.

118. Mainwaring, P.J., and Naylor, M.N.: A four-year clinical study to determine the caries-inhibiting effect of calcium glycerophosphate and sodium fluoride in calcium carbonate base dentifrices containing sodium monofluorophosphate. Caries Res., *17*:267–276, 1983.

119. Forsman, B.: Studies on the effect of dentifrices with low fluoride content. Community Dent. Oral Epidemiol., *2*:166–175, 1974.

120. Naylor, M.N. and Glass, R.L.: A 3-year clinical trial of calcium carbonate dentifrice containing calcium glycerophosphate and sodium monofluorophosphate. Caries Res., *13*:39–46, 1979.

121. Glass, R.L.: Caries reduction by a dentifrice containing sodium monofluorophosphate in a calcium carbonate. Partial explanation for diminishing caries prevalence. Clin. Prev. Dent., *3*:6–8, 1981.

122. Howat, A.P., Holloway, P.J., and Davies, T.G.H.: Caries prevention by daily supervised use of a MFP gel dentifrice. Report of a 3-year clinical trial. Br. Dent. J., *145*:233–235, 1978.

123. Rule, J., et al.: Anticaries properties of a 0.78% sodium monofluorophosphate-silica base dentifrice. Int. Assn. Dent. Res., Preprinted Abs. #921, 1982.

124. Autia, F.E., Shahni, D.R., and Kapadia, J.D.: A clinical study of sodium monofluorophosphate dentifrice in institutionalized children in the city of Bombay. J. Indian Dent. Assoc., *46*:165–170, 1974 (Oral Res. Abs., *10*:407, 1975).

125. Riethe, V.P., and Schubring, G.: Klinische Prufung einer Natriummonofluorphosphat-zahnpaste an Schulkindern. Dtsch. Zahnaerztl. Z., *30*:513–517, 1975.

126. Kobylanska, M.: Evaluation of the caries-inhibiting effect of Polish manufactured fluoride dentifrices. Czas. Stomatol., *28*:247–251, 1975.

127. Wilson, C.J., Triol, C.W., and Volpe, A.R.: The clinical anticaries effect of a fluoride dentifrice and mouthrinse. Int. Assn. Dent. Res., Preprinted Abs. #808, 1978.

128. American Dental Association, Council on Dental Therapeutics: Council classifies Colgate with MFP (sodium monofluorophosphate) in Group A. J. Am. Dent. Assoc., *79*:937–938, 1969.

129. American Dental Association, Council on Dental Therapeutics: Council accepts Macleans fluoride dentifrice. J. Am. Dent. Assoc., *92*:966–967, 1976.

130. American Dental Association, Council on Dental Therapeutics: Council accepts Aquafresh. J. Am. Dent. Assoc., *97*:80, 1978.

131. American Dental Association, Council on Dental Therapeutics: Council accepts Aim with sodium monofluorophosphate. J. Am. Dent. Assoc., *101*:822, 1980.

132. Triol, C.W., Wilson, C.J., and Volpe, A.R.: Effect on caries of two monofluorophosphate dentifrices in a nonfluoridated water area: A thirty-one-month study. Clin. Prev. Dent., *3*:5–7, 1981.

133. American Dental Association: Dentifrices which contain sodium monofluorophosphate. Accepted Dental Therapeutics, 39th Ed. 1982, pp. 358–359.

134. Mitropoulos, C.M., Davies, T.G.H., and Worthington, H.V.: Clinical comparison of two dentifrices with different levels of sodium monofluorophosphate. Int. Assn. Dent. Res., Preprinted Abs. #70:543, 1982.

135. Hanachowicz, L.: Caries prevention using a 1.2% sodium monofluorophosphate dentifrice in an aluminium oxide trihydrate base. Community Dent. Oral Epidemiol., *12*:10–16, 1984.

136. Hargreaves, J.A., and Chester, C.G.: Clinical trial among Scottish children of an anti-caries dentifrice containing 2% sodium monofluorophosphate. Community Dent. Oral Epidemiol., *1*:47–57, 1973.

137. Lind, O.P., Møller, I.J., von der Fehr, F.R., and Joost Larsen, M.: Caries-preventive effect of a

dentifrice containing 2% sodium monofluorophosphate in a natural fluoride area in Denmark. Community Dent. Oral Epidemiol., *2*:104–113, 1974.

138. James, P.M.C., Anderson, R.J., Beal, J.F., and Bradnock, G.: A 3-year clinical trial of a dentifrice containing 2% sodium monofluorophosphate. Community Dent. Oral Epidemiol., *5*:67–72, 1977.

139. Jiraskova, M., et al.: Effect of Czechoslovak-made toothpaste containing sodium fluoride. Cesk. Stomatol., *65*:433–436, 1965.

140. Wellock, W.D., and Brudevold, F.: A study of acidulated fluoride solutions. II. The caries inhibiting effect of single annual topical applications of an acidic fluoride and phosphate solution. A two-year experience. Arch. Oral Biol., *8*:179–182, 1963.

141. Wellock, W.D., Maitland, A., and Brudevold, F.: Caries increments, tooth discoloration, and state of oral hygiene in children given single annual applications of acid phosphate fluoride and stannous fluoride. Arch. Oral Biol., *10*:453–460, 1965.

142. Gutherz, M.: Klinischer Nachweis der Karieshemmwirkung einer metaphosphathaltigen Fluorzahnpaste. Schweiz. Mschr. Zahnheilk., *78*:235–247, 1968.

143. Reed, M.W., and King, J.D.: A clinical evaluation of a sodium fluoride dentifrice. Int. Assn. Dent. Res., Preprinted Abs. #340:133, 1970.

144. Weisenstein, P.R., and Zacherl, W.A.: A multiple-examiner clinical evaluation of a sodium fluoride dentifrice. J. Am. Dent. Assoc., *84*:621–623, 1972.

145. Reed, M.W.: Clinical evaluation of three concentrations of sodium fluoride in dentifrices. J. Am. Dent. Assoc., *80*:1401–1403, 1973.

146. Stookey, G.K., and Beiswanger, B.B.: Influence of an experimental sodium fluoride dentifrice on dental caries incidence in children. J. Dent. Res., *54*:53–58, 1975.

147. Ennever, J., et al.: Influence of alkaline pH on the effectiveness of sodium fluoride dentifrices. J. Dent. Res., *59*:658–661, 1980.

148. Approved new drug application for Gleem II toothpaste, NDA #16-985, Federal Register, October 29, 1973.

149. Valery, P., and Prevot, H.: Clinical results of the use of a fluoridated dentifrice. Chir. Dent. France, *41*:21–23, 1971.

150. Onisi, M.: The caries preventive effect of a new dentifrice containing NaF and synthetic zeolite. Jpn. J. Dent. Health, *24*:1–5, 1974.

151. Koch, G.: Effect of sodium fluoride in dentifrice and mouthwash on incidence of dental caries in school children. 8. Effect of daily supervised toothbrushing with a sodium fluoride dentifrice. A 3-year double-blind clinical test. Odont. Revy., *18*(Suppl. 12):1–125, 1967.

152. Koch, G.: Effect of sodium fluoride in dentifrice and mouthwash on incidence of dental caries in school children. 9. Effect of unsupervised toothbrushing at home with a sodium fluoride dentifrice. A 2-year double-blind clinical test. Odont. Revy., *18*(Suppl. 12):1–125, 1967.

153. Koch, G.: Selection and caries prophylaxis of children with high caries activity. One-year results. Odont. Revy. (Malmo), *21*:71–82, 1970.

154. Beiswanger, B.B., Gish, C.W., and Mallatt, M.E.: A three-year study of the effect of a sodium fluoride-silica abrasive dentifrice on dental caries. Pharmacol. Ther. Dent., *6*:9–16, 1981.

155. Edlund, D., and Koch, G.: Effect on caries of daily supervised toothbrushing with sodium monofluorophosphate and sodium fluoride dentifrices after 3 years. Scand. J. Dent. Res., *85*:41–45, 1977.

156. Edward, S., and Torell, P.: Constituents of dentifrices. Abstracts of Papers, 24th ORCA Congress, Caries Res., *12*:107, 1978.

157. Koch, G., Petersson, L.G., Kling, E., and Kling, L.: Effect of 250 and 1000 ppm fluoride dentifrice on caries. A three-year clinical study. Swed. Dent. J., *6*:233–238, 1982.

158. Silverstone, L.M.: Remineralization phenomena. Caries Res., *11*(Suppl. 1):59–84, 1977.

159. Kidd, E.A.M.: The histopathology of enamel caries in young and old permanent teeth. Br. Dent. J., *155*:196–198, 1983.

160. Silverstone, L.M.: The effect of fluoride in the remineralization of enamel caries and caries-like lesions in vitro. J. Public Health Dent., *42*:42–53, 1982.

161. ten Cate, J.M., and Arends, J.: Remineralization of artificial enamel lesions in vitro. III. A study of the deposition mechanism. Caries Res., *14*:351–358, 1980.

162. Featherstone, J.D.B., Cutress, T.W., Rodgers, B.E., and Dennison, P.J.: Remineralization of artificial caries-like lesions in vivo by a self-administered mouthrinse or paste. Caries Res., *16*:235–242, 1982.

163. Cilley, W.A., and Haberman, J.P.: Fluoride in enamel and correlation to caries. Int. Assn. Dent. Res., Preprinted Abs. #1069, 1981.

164. Mobley, M.J.: Fluoride uptake from in situ brushing with a SnF_2 and a NaF dentifrice. J. Dent. Res., *60*:1943–1948, 1981.

165. Stookey, G.K., et al.: In situ fluoride uptake from fluoride dentifrices. Int. Assn. Dent. Res., Preprinted Abs. #118, 1984.

166. Mellberg, J.R., and Chomicki, W.G.: Effect of soluble calcium on fluoride uptake by enamel from sodium monofluorophosphate. Int. Assn. Dent. Res., Preprinted Abs. #281, 1982.

167. Mellberg, J.R., and Chomicki, W.G.: Fluoride uptake by artificial caries lesions from fluoride dentifrices in vivo. J. Dent. Res., *62*:540–542, 1983.

168. Mellberg, J.R., and Chomicki, W.G.: Calcium effect on fluoride uptake from monofluorophosphate in vivo. Int. Assn. Dent. Res., Preprinted Abs. #764, 1984.

169. Mallon, D.E., and Mellberg, J.R.: Calcium enhanced in vitro remineralization by monofluorophosphate. Am. Assn. Dent. Res., Preprinted Abs. #475, 1983.

170. Mellberg, J.R., Chomicki, W.G., Mallon, D.E., and Castrovince, L.A.: In vivo remineralization of artificial caries lesions by an MFP dentifrice. Int. Assn. Dent. Res., Preprinted Abs. #763, 1984.

171. White, D.J., Lueders, R.A., and Mobley, M.J.: A comparison of in vitro remineralization from fluoride dentifrices. Int. Assn. Dent. Res., Preprinted Abs. #624, 1984.

172. Park, K.K., Wood, G.D., Schemehorn, B.R., and Stookey, G.K.: Remineralization from saliva and fluoride dentifrices. Int. Assn. Dent. Res., Preprinted Abs. #119, 1984.

173. Mellberg, J.R.: Monofluorophosphate utilization in oral preparations: Laboratory observations. Caries Res., *17*(Suppl. 1):102–118, 1983.

174. Featherstone, J.D.B., ten Cate, J.M., Shariati, M., and Arends, J.: Comparison of artificial caries-like lesions by quantitative microradiography and microhardness profiles. Caries Res., *17*:385–391, 1983.

175. Gerdin, P.O., and Serneke, D.: Studies in Dentifrices, VII: Fluoride dentifrices and remineralization of dental enamel surfaces. An in vivo study. Swed. Dent. J., *66*:249–270, 1973.

Chapter 10
LABORATORY METHODS OF ASSESSING FLUORIDE DENTIFRICES AND OTHER TOPICAL FLUORIDE AGENTS

Conrad A. Naleway

The Laboratory of the Council on Dental Therapeutics has the responsibility of assisting the Council in its review of products for the Acceptance Program. This responsibility includes evaluating the chemical integrity of all products under review for acceptance by the Council and monitoring the consistency of those products that have been accepted. Among those products evaluated by the Council are dentifrices and professional topical fluoride products such as acidulated phosphate fluoride (APF) gels and stannous fluoride (SnF_2) products. In its review of these products, the Laboratory will periodically compare all accepted products of a certain defined classification utilizing specific chemical or in vitro models. These comparative studies permit the Council and its staff to better view the composition of commercial products using state-of-the-art technologies and to learn from as well as contribute to an understanding of how these agents and products function. This work has proven to be useful in the development and the anticipated implementation of laboratory guidelines for the acceptance of generic dentifrices. This chapter also reviews the current work in the evaluation of (1) APF products; (2) 0.4% SnF_2 gels; and (3) effects of topically applied fluoride products on surface alteration of porcelain materials.

One of the primary responsibilities of the Council on Dental Therapeutics is the evaluation of dentally related drugs to determine their therapeutic efficacy. An important aspect of this Acceptance Program is the review of clinical studies submitted by manufacturers to support the efficacy of their product(s). The Council utilizes consultants in various fields of expertise to assist in the review of submitted data. The Laboratory of the Council on Dental Therapeutics has the responsibility to assist the Council in this review mechanism. This responsibility includes reviewing the chemical integrity of all products under review for acceptance by the Council and monitoring the consistency of those products that have been accepted. Among those products evaluated by the Council are the OTC dentifrices and professional topical fluoride products such as acidulated phosphate fluoride gels and stannous fluoride products.

In its review of these products, the Laboratory will periodically compare all accepted products of a certain defined classification utilizing specific chemical or in vitro models. Such comparative studies permit the Council and its staff to better view the composition of commercial products using state-of-the-art technologies and to learn from as well as contribute to an understanding of how these agents and products function. This work has proven to be very useful in cases such as the development and the anticipated implementation of laboratory guidelines for the acceptance of generic dentifrices. This chapter will attempt to present an overview of the types of research activities that have been undertaken by the Laboratory over the past few years. It is not my intent for the overview to be totally comprehensive but rather to address the interrelationship of four specific laboratory programs with both the Acceptance Program and the clinical efficacy of a variety of fluoride agents.

COUNCIL REVIEW OF FLUORIDE DENTIFRICES

One of the most recent reviews in which the Council has been involved and at the present time is still deliberating is the role of laboratory studies in the assessment of the clinical efficacy of fluoride dentifrices. All therapeutic dentifrices that are currently accepted have been evaluated based upon a review of well-controlled clinical studies submitted by manufacturers. Table 10–1 lists these products. These studies have been conducted with the same basic formulation as the currently manufactured product. This approach by the Council (i.e., requiring clinical trials for each dentifrice submission), has been influenced by a number of factors:

(1) There exists a concern within the scientific community regarding the complexity of dentifrice formulations. Information collected by some researchers suggests that the type and physical character of the abrasive, as well as other soluble components (including fluoride), affect the chemical and biochemical activity of the finished product.

(2) A review of the clinical literature suggests a wide range of results regarding the clinical efficacy of similarly formulated dentifrices. A number of unanswered questions exist concerning the chemical composition of many of these tested formulations. The relationship, in general, between any specific chemical indicator of possible clinical effectiveness (e.g., rat caries) and the relative efficacy of a specific formulation is weak.

(3) The consumer recognizes the Association Seal as a reliable indicator of product efficacy. When evaluating products purchased by consumers on an over-the-counter basis, the Council has based its evaluation on sound scientific data and required that manufacturers limit claims for the products and their place in an oral hygiene program to that supported by the data, avoiding speculative or nonsupportable positions. Using these methods, the Association has acted deliberately to avoid recognizing or accepting ineffective or inactive products.

Table 10–1. Accepted Fluoride Dentifrices

AIM Toothpaste, Mint, Regular—Lever Brothers Co.
AQUA-FRESH—Beecham Products
COLGATE GEL Toothpaste with MFP—Colgate-Palmolive Co.
COLGATE with MFP Fluoride—Colgate-Palmolive Co.
CREST Toothpaste, Mint, Regular—Procter & Gamble
GEL FORMULA CREST—Procter & Gamble
MACLEANS Fluoride Toothpaste, Peppermint, Regular—Beecham Products

During the past 18 months the Council and its staff have actively reviewed their position with leading fluoride and dentifrice researchers. The question that has been addressed is, "Based upon our understanding of fluoride activity as it relates to dentifrices, can the clinical efficacy of fluoride dentifrices be determined by a set of laboratory profiles?"

As a result of these discussions, the Council requested that its staff draft an appropriate set of guidelines for reviewing all dentifrices that would be scientifically supportable, legally sound, and technically workable for both ADA staff and manufacturers.

Up to this time, the Council has required clinical trials to support the efficacy for all new therapeutics dentifrices as well as currently accepted dentifrices which have undergone substantial changes in formulation. A product formulation was defined as new if the active fluoride agent and/or abrasive system was not previously tested or was clinically tested but found not to be clinically effective. Furthermore, substantial changes in either the nature or level of the fluoride/abrasive system also required independent clinical testing to confirm efficacy. Alterations in a dentifrice which modified such factors as the coloring, flavoring, and humectant, as well as small adjustments in the percentages of abrasive(s), have not generally required additional clinical studies. Laboratory and animal caries data, however, were requested in some instances to support the equivalence of the modified formulation with the previously clinically tested formulation.

In general, unwritten guidelines have been utilized in the past in the acceptance review of fluoride dentifrices as well as all other OTC products. Due to the new philosophy that will permit a specific class of product (e.g., fluoride dentifrices) to be approved based solely upon laboratory data and the complexity associated with defining a comprehensive collection of laboratory indicators, it is now necessary to formalize the Council position.

Council's Proposed Guidelines for the Acceptance of Fluoride Dentifrices

It is proposed that the following criteria be utilized by the Council in its review of fluoride dentifrices:

(1) The Council's review and acceptance of fluoride dentifrices will be based upon a scientific review of laboratory and/or clinical data on each submitted product. This acceptance status will remain essentially the same as that presently used to recognize accepted dentifrice products. Products accepted in this program could use the Seal of Acceptance, a statement relative to the use of the product, and appropriate claims and/or notation, which is scientifically supportable, in advertising and promotional material. For example, products accepted on the basis of clinical studies would be able to refer specifically to those studies.

(2) The acceptance process will be maintained relatively unchanged. The scientific data submitted by the manufacturer will still need to address the question of clinical efficacy and the safety of each product. *Laboratory, animal, and in situ experimentation in support of a product's equivalency to a previously tested, clinically effective product(s), however, may be more widely utilized for some products.* The required testing protocol will need to be comprehensive and reflect *current methodology, which is appropriate at time of submission,* in evaluating the potential activity of a fluoride dentifrice. More specifically:

(A) Clinical studies will still be required for all *new* fluoride/abrasive systems.

(B) Clinical studies will still be encouraged since they permit the only definitive evidence of the efficacy of a dentifrice. An alternative approach may be used at the option of each manufacturer. Formulations that are able to demonstrate chemical and biochemical equivalence with previously clinically tested products will be reviewed by the guidelines defined in the section, *Required Materials Needed for Acceptance as Generic Equivalent*.

(C) It will be the obligation of the manufacturer to perform all tests and to assemble the scientific material necessary to support the equivalence of the new formulation with formulations that have been previously clinically tested. Council staff, however, will be available to discuss research protocol for implementation of the study.

(D) The Council will continue to request assistance from its scientific consultants in its review of *all* dentifrice submissions. Each completed submission will be reviewed critically by independent Council consultants from the scientific community. This phase of the review process is an integral component of the Acceptance Program and has been consistently utilized in the review of all clinical studies.

(E) Required laboratory data necessary to support the efficacy of a dentifrice will be determined by the specific formulation. Potential chemical incompatibilities and/or differences of the test product from the originally clinically tested formulation(s) will serve as a basis for evaluation. This will include a review of the physical, chemical, and biochemical characteristics of the test product and the possibility that such differences may influence the bioactivity of the fluoride agent. A determination of the required laboratory data necessary will be coordinated with the manufacturer.

(F) These guidelines will be utilized in the review of all fluoride dentifrice formulations. In those specific product formulations where ingredients have been introduced to elicit additional therapeutic or cosmetic effects, the manufacturer will be required to submit supporting documentation to prove such effects in addition to the data defined in these guidelines.

Required Material Needed for Acceptance as Generic Equivalent

The following acceptance process will be used in evaluating fluoride dentifrices for the Acceptance Program. Each product will be required to submit supporting data in each of the following five categories:

(1) *ANIMAL CARIES STUDIES*

The animal model incorporates to a substantial degree the collective effects of the treatment product (1) to deliver fluoride to the enamel; (2) to effect both remineralization and demineralization; and (3) to potentially alter the bacterial flora within the plaque upon the tooth structure. The design of the animal study is flexible, but must be scientifically sound and defendable.

(2) *FLUORIDE AVAILABILITY AND STABILITY*

Chemical data must be submitted to document that the active fluoride agent is chemically free and available in both fresh and aged samples.

(3) *FLUORIDE BIOAVAILABILITY IN ENAMEL*

Each test dentifrice must demonstrate an ability to deliver and incorporate levels of fluoride into both sound and decalcified enamel, which are equivalent with clinically tested formulation(s).

(4) *ABILITY OF PRODUCT TO ENHANCE REMINERALIZATION*

The experimental protocol must demonstrate quantitatively that the test product remineralizes tooth enamel to the same degree as similar clinically tested formulations. Remineralization of decalcified enamel may be simulated in either an in situ or in vitro environment.

(5) *ABILITY OF THE PRODUCT TO DIMINISH DEMINERALIZATION*

The test product must be able to diminish enamel demineralization to the same degree as similar clinically tested formulations. Decalcification of the tooth structure must be determined in a controlled environment where the chemical and/or biochemical conditions approximating the mouth need to be simulated.

When changes in formulation are related to changes that do not affect the fluoride activity (e.g., changes in coloring or flavor), specific tests may be waived if the manufacturer is able to present adequate rationale for not requiring such tests. Written documentation should be presented in support of tests being waived. Supporting data for fluoride availability and stability, animal caries, and enamel fluoride bioavailability, however, will be required for all submissions.

The design of each profile study must be in accordance with existing recognized research methodology. Minimally, it must contain: (1) the appropriate placebo (test product minus active fluoride agent); (2) test product in formulation equivalent to that to be manufactured; (3) the clinically tested formulation and compositional equivalent containing the same active agent/abrasive system, and; (4) (optional but recommended), Council suggests that more than one clinically tested product be included in each laboratory profile study in order to define the degree of variance associated with each testing profile. The strongest possible case for potential efficacy would be based upon the test formulation being a part of the statistical subgroup of known therapeutically active products all of which are found to be statistically different from placebo.

The statistical analysis of each study must clearly separate the test product from the placebo and strongly suggest that the test formulation is equivalent to the clinically tested formulation(s).

QUALIFIED COUNCIL RECOGNITION OF COMPATIBLE FLUORIDE DENTIFRICES

A new class of Council recognition is under consideration. One of its primary goals would be to assist the consumer in differentiating potentially active from potentially nonactive dentifrice products. It will also attempt to identify dentifrices that are chemically stable and consistent. The limited battery of tests associated with this class of product will be either conducted in cooperation with the manufacturer, in which case the manufacturer will be able to advertise such recognition, or testing may be performed in our laboratories and/or in cooperation with other academic laboratories resulting in the periodic joint publication of both positive and negative findings. Both advertising and labeling would need to reflect the lesser degree of scientific supporting data, which has been made available concerning these specific products. All advertising copy directly associated with the use of our limited recognition will require review by the Council. In order to warrant recognition by the Council, each product will need to conform to the following minimal standards. It must:

(1) Contain a fluoride/abrasive system which has been clinically demonstrated to be effective.

(2) Contain available free and total fluoride as defined by the FDA's Proprietary Committee report.

(3) Demonstrate acceptable stability over the shelf life of the product.

(4) Contain no clear or suspect chemical inconsistency that might suggest need for further testing to warrant proof of bioavailability. It will be the responsibility of the manufacturer to present sufficient data to convince both staff and Council consultants that the product is potentially active.

(5) Maintain a high level of quality control. Tests on randomly selected fresh samples will need to contain labeled levels plus or minus 10%.

COUNCIL REVIEW OF APF GELS

The topical application of sodium fluoride solutions and gels acidified with orthophosphoric acid has been demonstrated by clinical studies to have a considerable caries-inhibiting effect in children.[1-3] Although all studies did not utilize exactly equivalent levels of phosphate and fluoride, and the clinical methodology differed, the reduction in caries incidence in each instance was statistically significant. The preparations usually employed in the clinical studies contained 1.23% F and approximately 1% orthophosphoric acid. As in the case of all clinical studies, variation in the degree of clinical efficacy differed from study to study. In the case of the preceding three studies, use of this preparation over 2 years or more resulted in a reduction in caries incidence (DMFS) varying from 26 to 70%. Acidulated phosphate fluoride gels have been accepted by the Council since 1968 with the first commercial product being Karidium Phosphate Fluoride Topical Gel manufactured by the Lorvic Corporation. Table 10–2 contains an up-to-date list of all accepted APF gels.

From 1979 to 1980 a survey of acidulated phosphate fluoride gels obtained from the open market was conducted by the Division of Chemistry. This survey found that nearly half of those products evaluated did not fall within U.S.P. specifications. Additionally, significant variability was found between different lots of the same product. Because of these variations the Council was concerned that dentists may not be receiving consistently high quality products for treating their patients.

At its June 1980 meeting the Council on Dental Therapeutics reiterated that all currently accepted acidulated phosphate fluoride gels must be tested and must meet the specifications outlined in the current *United States Pharmacopeia* (USP) before they can be considered for continued acceptance. Manufacturers were notified that it was their responsibility to maintain a quality control program and to insure that all formulations maintain consistency in chemical composition and physical properties.

As a result, the Council has indicated that manufacturers must submit evidence that their products meet USP specifications for pH, viscosity, and fluoride prior to initial acceptance as new products or during reconsideration of currently accepted products. Additionally, the Council indicated that all technical claims such as thixotropy must be supported by appropriate supplementary studies, which are acceptable to the Division of Chemistry.

The Division of Chemistry was requested to conduct independent tests on samples submitted by manufacturers prior to acceptance or reconsideration of products. In addition, spot checks were to be made periodically on samples obtained from ran-

Table 10–2.   Accepted Topical APF Preparations

Butler Topical Fluoride Phosphate Anticaries Gel—John O. Butler Co.
Centra Guardian Angel Topical Fluoride Gel—Centra Dental Products
Fluorident Gel—Premier Dental Products Co.
Fluorident Liquid—Premier Dental Products Co.
Gell II—CooperCare, Inc.
Gelution Topical Fluoride—Unitek Corp.
Healthco Fluoride Gel, VM—Healthco, Inc.
Healthco Topical Fluoride Gel—Healthco, Inc.
Iradicav Acidulated Phosphate Fluoride Solution—Janar Co., Inc.
Janar's Acidulated Phosphate Fluoride Rinse—Janar Co., Inc.
Karidium Phosphate Fluoride Topical Gel—Lorvic Corp.
Karidium Phosphate Fluoride Topical Solution—Lorvic Corp.
Karidium Thixotropic Acidulated Phosphate Fluoride Gel—Lorvic Corp.
Kerr Fluoride Gel—Sybron/Kerr
Luride Topical Gel—Hoyt Laboratories
Luride Topical Solution—Hoyt Laboratories
Nufluor Acidulated Phosphate Fluoride Topical Gel—Janar Co., Inc.
Pacemaker Topical Fluoride Gel—CooperCare, Inc.
Pacemaker Topical Fluoride Solution—CooperCare, Inc.
Pennwhite Topical Fluoride Gel—S.S. White Retail
Phos-Flur Oral Rinse Supplement—Hoyt Laboratories
Predent Topical Fluoride Treatment Gel—Harry J. Bosworth Co.
Rafluor New Age Gel—Pascal Co., Inc.
Rafluor Topical Gel—Pascal Co., Inc.
Rafluor Topical Solution—Pascal Co., Inc.
Sabragel—Sabra Dental Products, Inc.
Sultan Topical Fluoride Gel—Sultan Chemists, Inc.
Super-dent Topical Fluoride Gel—Rugby Laboratories, Inc.
Thera-Flur (acidulated) Gel-Drops—Hoyt Laboratories
Thixo-Flur Thixotropic Topical Gel—Hoyt Laboratories
Topical Fluoride Gel—Professional Way Corp.

domly selected dentists and/or distributors throughout a manufacturer's distribution region. As a result, all fluoride products are now carefully screened before acceptance. Samples of products that have been reviewed by the Council since that time have been found to maintain the required USP characteristics.

Because of the great diversity of results obtained with the various APF products tested, our Laboratory undertook an in vitro study to examine how such diversity of products could affect the chemical and/or biochemical activity of the product. The following results address this point.

Comparative In Vitro Study of Commercial APF Gels

The therapeutic benefit of acidulated phosphate fluoride gel treatment has been demonstrated clinically. The actual mechanism by which these gels operate, however, is not fully understood. Although a fraction of the retained fluoride following APF treatment is incorporated into enamel in an apatitic form, the majority of the fluoride is present as calcium fluoride (CaF_2). Calcium fluoride is a rather soluble salt which is produced at the expense of the enamel surface.[4–6] It has been proposed that calcium fluoride may reduce enamel solubility by acting as a source of fluoride in the slow formation of fluorapatite.[7] It is further suggested that these topically applied fluorides may affect the surface character of enamel, thus altering its microbial colonization,[8,9] or, the slow release of fluoride as a result of the dissolution of calcium fluoride may inhibit bacterial glycolysis.[10]

This study was undertaken in an attempt to comprehensively evaluate the physical, chemical, and biochemical properties of the various APF gels. Our study was strongly influenced by the work of Zahradnik et al.[11] and the private discussions with coauthor Moreno.[11] Their study strongly suggested that the reduction in the degree of demineralization of APF-treated enamel using a chemical agent (lactic acid) was small. The reduction in the degree of demineralization due to a bacterial challenge following APF treatments as compared with the control treatment, however, was substantial.

Using an in vitro model, 19 APF gels were compared under two experimental approaches. One approach examined the total fluoride incorporation into enamel. The second measured the extent of enamel dissolution to a challenge of acids produced by micro-organisms. The selection of these gels was based upon their acceptance by the Council on Dental Therapeutics from 1979 to 1980.

Method.   The enamel substrate used was four bovine maxillary incisors (i.e., the centrals and laterals). The bovine teeth were cleaned, pumiced, and then air dried. The labial surface of each tooth was covered with nail polish except for three windows. Two of these windows were used for the determination of fluoride uptake, and the third window for enamel dissolution.

Each tooth was treated with the APF gel by applying the gel directly to a window for a period of 4 min. They were then rinsed with deionized water and the excess was wiped off with a moist chamois. The teeth were then immersed in an artificial inorganic saliva for 1 hour and then transferred to another solution of the artificial saliva for an additional 23 hours. This fluid simulated human saliva, but contained no fluoride. The saliva solutions were analyzed for fluoride content. The teeth were then immersed in 1.5 ml of 0.5 N $HClO_4$ for 30, 40, 50, and 60 sec respectively. 1.5 ml of adjusted TISAB was added after each immersion in the perchloric acid. Analysis was then made for calcium, fluoride, and phosphate at each of the four layers. Fluoride was analyzed using a specific ion electrode; calcium was analyzed with an atomic absorption spectrophotometer; and phosphate was analyzed by forming the standard molybdate complex.

The enamel dissolution experiments employed the *Streptococcus mutans* strain 6715. This strain produces more insoluble glucans than many other strains and also adheres better to the enamel surface. The third window on the labial surface of the 200 bovine teeth utilized in this study was exposed. The teeth were again treated with an APF gel for 4 min, rinsed with deionized water and the excess wiped off with a moist chamois. The teeth were then immersed in artificial saliva for a total of 24 hours in two steps as discussed earlier. They were then exposed for 3 days in heart infusion broth containing sucrose innoculated with 0.1 ml of freshly cultured *Streptococcus mutans* in an incubator under anaerobic conditions. The teeth were then transferred to fresh heart infusion broth for a further 3 days. The broths were analyzed for pH and total calcium. Sterile conditions were maintained throughout.

Results.   A cubic spline was fitted to the four cumulative fluoride uptake totals. This function was then differentiated to obtain the interpolated fluoride concentration at comparable depths. This technique is advantageous in that the fitting function is consistent with all the experimental data.

Since the fluoride uptake and subsequent enamel resistance to dissolution were expected to vary from jaw to jaw, a balanced incomplete block design was used to analyze the data. A total of 25 treatments (including two nonfluoride controls) were compared and assigned to each tooth such that each treatment was applied eight times and every possible pair of treatments occurred once within a jaw. This design permits all comparisons among pairs of treatments to be made with equal precision.

As much as two-thirds of the total fluoride absorbed by the bovine enamel was released into the artificial saliva in the 24-hour period following treatment. This is presented in Table 10–3. Between 25 to 65% of the fluoride released into the saliva was lost within the first hour. The analysis of fluoride uptake absorbed into the enamel does not include this fluoride release into the saliva.

Total fluoride uptake was determined at median depths of 7.6, 18.1, 30.3, and 45.7 microns, respectively, on the bovine teeth. Fluoride concentrations were estimated at depths of 2, 5, and 17 microns in order to compare the APF gel treatment. Mean fluoride levels which have been adjusted by the balanced incomplete block design are presented in Figure 10–1.

Table 10–4 presents an ANOVA breakdown for the fluoride uptake interpolated at a depth of 5 microns. The results were essentially the same at each depth. The level of fluoride uptake in bovine enamel is significantly different among the various APF gel treatments. Although these differences were significant at the three depths examined, the magnitude of the differences diminished at each succeeding depth.

Because of both the confidentiality which exists between manufacturers and the ADA and the lack of clear correlation between the results of these studies and the clinical efficacy of commercial products, only the chemical characteristic of products

Table 10–3. Fluoride Incorporation into Bovine Enamel

Treatment #	pH	Integrated Mean Fluoride Uptake*	Fluoride Lost to Artificial Saliva		
			1 hr	23 hrs	Total
16	2.20	1.06	1.10	0.65	2.81
10	3.10	0.76	0.76	0.50	2.02
24	3.15	3.67	1.38	1.99	7.04
14	3.20	4.47	1.07	2.11	7.65
22	3.20	1.87	0.69	1.68	4.24
8	3.25	1.74	0.68	0.63	3.04
13	3.25	1.92	0.89	0.67	3.48
12	3.25	1.39	0.98	0.83	3.20
21	3.30	1.11	0.51	0.46	2.08
15	3.30	2.30	1.39	1.25	4.94
7	3.40	0.34	0.14	0.16	0.64
3	3.40	0.31	0.19	0.14	0.64
1	3.40	0.35	0.37	0.18	0.89
18	3.45	1.78	0.84	1.13	3.75
19	3.60	2.10	0.99	0.98	3.98
6	3.90	2.74	0.71	0.70	4.15
20	4.15	2.30	1.33	1.18	4.81
4	4.20	0.88	0.47	0.53	1.88
25	4.25	1.34	0.47	1.58	3.39
23	4.30	3.01	0.96	1.78	5.75
11	4.30	1.54	0.63	2.04	4.21
17	4.80	2.07	0.55	0.54	3.16
5	5.67	0.54	0.35	0.20	1.09
9 (Control)	—	0.19	0.18	0.22	0.59
2 (Control)	—	0.15	0.26	0.16	0.57

*Micrograms of fluoride

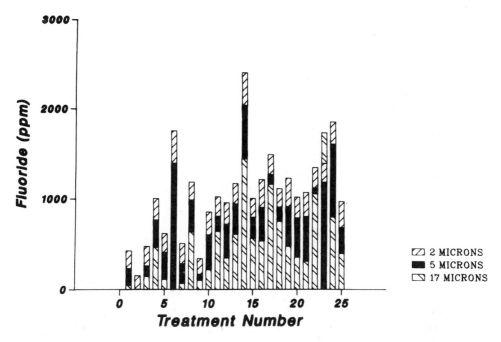

Fig. 10–1. Fluoride uptake in bovine enamel following topical treatment. Means adjusted for balanced incomplete block design. Fluoride levels interpolated at depths of 2, 5, and 17 microns are presented for each treatment.

will be identified and not the product names. It is clearly not the intention of the Council to rank products.

The adjusted treatment means varied from no uptake for the controls to a range of 400 to 2400 ppm at 2 microns depth, 200 to 2000 ppm at 5 microns and 100 to 1700 ppm at 20 microns respectively. Except for minor changes, the ordering of the treatments did not alter up to a depth of 20 microns. Furthermore, there was no clear relationship between the fluoride uptake in the bovine enamel and the pH level of the APF gel.

The spread in degree of fluoride uptake into bovine enamel as presented in Figure 10–1 permits statistical separation between treatments to be defined. There were statistical differences between products with the highest degree of fluoride incorporation and the control groups with the two lowest levels of detected fluoride. Furthermore, a number of commercial products were not statistically different from the control in fluoride uptake. Coding of treatments is consistently defined throughout

Table 10–4. ANOVA Incomplete Block Design on Fluoride Uptake Study

Source	D.F.	Mean Source	E
Treatment (Adj)	24	1.768	2.67*
Block (Unadj)	49	0.949	1.44
Error	126	0.658	
Total	199		

*Significant at the 1.0% level

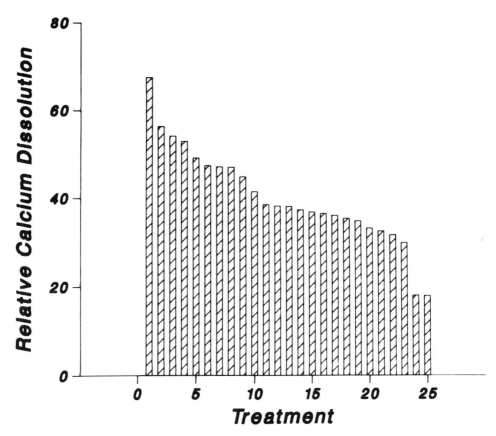

Fig. 10–2. Enamel decalcification in treated bovine enamel using *Streptococcus mutans* strain 6715. Means adjusted for model design. Calcium levels are corrected for surface area and solution volume. Model design attempts to account for variation in microbiological variation by week and differences between bovine jaws.

this section by the experimental ordering as determined by the microbiological study presented in Figure 10–2.

Table 10–5 presents the ANOVA results for the microbiological study which monitored the degree of enamel dissolution in bovine teeth following various fluoride treatments. It should be noted that the model used to analyze the dissolution data attempted to account for the observed differences in bacterial growth from week to

Table 10–5. ANOVA Incomplete Block Design on Streptococcus mutans Dissolution Study

Source	D.F.	Mean Square	F
Week (Unadj)	5	102,265	286.72*
Block (Adj)	44	6,955	19.50*
Treatment (Adj)	24	1,316	3.63*
Week* Treatment	90	1,607	4.51*
Error	35	356.7	
Total	198		

*Significant at the 1.0% level

week. This statistical adjunct does not substantially alter the ordering of treatments but does dramatically affect the level of statistical separability between treatments.

The level of calcium dissolution from bovine enamel challenged with *Streptococcus mutans* is significantly different between the various fluoride treatments. These differences were found to be significant during the first 3 days, the second period of 3 days and the total 6-day period. The degree of dissolution during the second 3-day period, however, was substantially less than during the first period. Again as in the case of fluoride uptake, there were products that were both statistically indistinguishable from the controls and separable from products with high resistance to dissolution (i.e., low dissolution rates). In contrast to the uptake study, the controls in the dissolution study were not at the extreme of the treatment group. Figure 10–2 summarizes the data adjusted by the incomplete block design for the first 3-day period; the control treatments are defined as 2 and 7.

Figure 10–3 displays the correlation between the mean results from the uptake study and the dissolution study. A correlation coefficient of (-0.563) was obtained, which indicated a level of statistical significance of $p < 0.001$. There was no statistical relationship determined between the two dependent variables (fluoride uptake or calcium dissolution) with the pH, fluoride content, or viscosity of the treatment products.

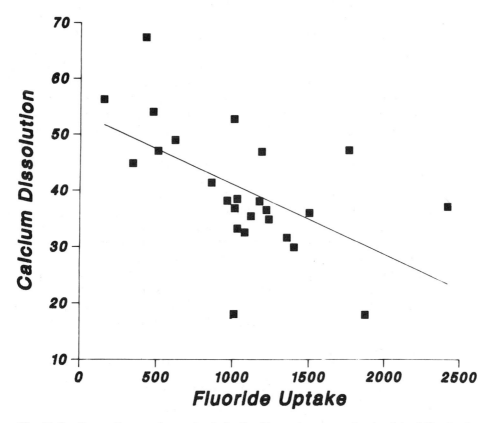

Fig. 10–3. Scatter diagram of mean levels for fluoride uptake vs mean levels of decalcification by microbiological challenge.

COUNCIL REVIEW OF STANNUS FLUORIDE GELS

0.4% stannous fluoride nonaqueous gels have been accepted by the Council since 1980 with the first accepted commercial product being Gel-Kam manufactured by Scherer Laboratories, Inc. A list of all currently accepted stannous fluoride gels is given in Table 10–6. These products with minor variations among manufacturers contain 0.4% by weight stannous fluoride, 98% glycerine, and a small quantity of flavoring, thickening, and preserving agents. 0.4% SnF_2 is equivalent to 970 ppm in F and thus is at the same level in fluoride concentration as most fluoride dentifrices. These products have been indicated for use once a day, preferably at night, and should be brushed on for 1 min.

0.4% stannous fluoride gels have been tested in clinical studies of a limited nature and are reported to be effective in preventing decalcification in orthodontic patients[12] and protecting against post-irradiation caries.[13] The acceptance of this class of product was further supported by the established efficacy of the former Crest stannous fluoride dentifrice which contained the same level of total fluoride.[14-15] Dr. Clifford Whall from our Council staff is in the process of drafting a Council Report on 0.4% stannous fluoride daily, home-use, brush-on gels. This document will further address the Council's position on this class of products.

As part of the Council on Dental Therapeutics' ongoing product analysis program, the Council's Division of Chemistry in 1982 examined samples of a number of 0.4% stannous fluoride gels for fluoride, total tin, and stannous ion content. Both the fluoride and total tin content of these products was found to fall within the 5 to 10% acceptable variation about the labeled level. Less than 30% of the tin, however, was detected as stannous ion using the iodate procedure of Hefferren.[16] The average levels for three of the products reviewed at that time are shown in Figure 10–4.

Since the total tin content was present as labeled, there exists a number of possibilities regarding the variation in the content of stannous ion. One possibility is that the testing procedure employed is not directly applicable to nonaqueous stannous fluoride gels. The iodometric determination of stannous ion is nonspecific. The assay depends upon the quantified redox titration of stannous ion by iodine to form iodide and stannic ion. This same procedure, however, has been utilized by Shannon in studies in which he monitored the stability of such formulations.[17] Since commercial formulations often contain such ingredients as thickening, coloring, and flavoring agents, there is a possibility that these factors may alter the chemical matrix of the test system by complexing in some manner with the stannous ion. A second possibility is that the stannous ion has been oxidized to stannic ion. It is speculated that this could occur during the initial phases of product preparation, during storage, or if the glycerin used in manufacturing the product is not sufficiently water-free. Thermodynamically, in the presence of oxygen and with the likely small amount of water present in these products, oxidation of stannous to stannic ion is inevitable.

Table 10–6. Accepted Topical Stannous Fluoride Preparations

EASYgel—Du-More, Inc.
FLO-GEL 0.4% Stannous Fluoride Gel—Sabra Dental Products, Inc.
GEL-KAM 0.4% Stannous Fluoride Gel—Scherer Laboratories, Inc.
GEL-TIN 0.4% Stannous Fluoride Gel—Young Dental Mfg. Co.
OMNI 0.4% Stannous Fluoride Gel—Dunhall Pharmaceuticals, Inc.
STOP 0.4% Stannous Fluoride Gel—CooperCare, Inc.

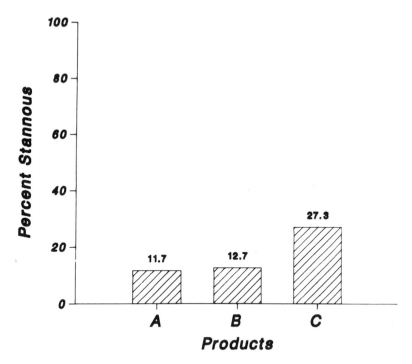

Fig. 10–4. Measured mean levels of free stannous ion in three commercial 0.4% stannous fluoride nonaqueous gels. Products were assayed in 1982. Percentages are defined in terms of labeled amount of stannous ion.

In order to examine the situation further, the Laboratory requested that manufacturers of all accepted products as well as products under review work with the Council in better defining and hopefully correcting this situation. Overall manufacturers were responsive in their participation with the Laboratory in its chemical review. As would be expected, however, there was a true spread in the degree of response from different manufacturers.

Manufacturers under the acceptance review were asked to submit two specimens each of five lots of varying age in their normal packaging containers. They were asked to retain samples from each of these lots for their own investigative review. We further asked that the following information be submitted: (1) a detailed outline of the manufacturing process; (2) documentation on studies regarding the air-tight quality of product containers; and (3) copies of any laboratory studies related to the chemical composition and stability of commercial products which might be appropriate.

As a general rule, the Council has required that if a product is to receive the Seal of Acceptance based upon clinical studies conducted using a chemically equivalent product, then this product must clearly demonstrate its **equivalence** to the clinically tested formulation. The commercial 0.4% stannous fluoride gels were accepted by the Council based on the clinical studies performed by Shannon et al., who used a stable nonaqueous gel. The laboratory study by Shannon mentioned earlier showed that the gel was chemically stable as 95.3% stannous fluoride for up to 15 months.[17]

At the January 1983 Council on Dental Therapeutics meeting, the Council reviewed its position on stannous fluoride gels. It was the Council's opinion that based upon a review of scientific literature concerning this product class, there is at the present time no clear relationship between the precise level of stannous ion in this class of product and its clinical efficacy. The Council, however, requested that the highest practical level of stability and consistency be maintained for all "accepted products." In order to maintain the highest practical standard the following supplementary information was transmitted to each manufacturer of an accepted product within this class.

(1) Manufacturers of accepted products will be requested to submit samples of the most recent lots of stabilized stannous fluoride product. Newly formulated batches of product should contain 90 to 105% of the labeled level of tin as stannous ion. Although there are various factors to be considered in the actual preparation of such a formulation, the following aspects should be addressed in maintaining the highest degree of stability:

(a) The highest practical grade of glycerol should be used. Both water and other trace contaminants may catalyze he oxidation of stannous to stannic ion.

(b) The minimum level of free oxygen should be permitted during both the mixing and packaging of formulation. This is especially true during periods when the product is raised to elevated temperatures.

(c) The stability and reactivity of any thickening agent should be examined.

(d) The concentration of stannous ion in both the raw material and the final product should be routinely determined to monitor possible deterioration or inconsistency of formulation.

(e) The highest grade of raw stannous fluoride product should be utilized.

(2) Both accelerated and ambient aging studies should be initiated on either lots as defined above in (1), or lots that have previously been determined to be stable as defined above. These should be submitted as they become available.

(3) Manufacturers were asked to introduce onto their labeling either an expiration date or a statement defining rate of anticipated degradation of the stannous ion in their product. Initially, this period or rate of alteration can be based upon either accelerated (high temperature) stability studies or other in-house ambient aging data which the manufacturers might have now or in the near future. It was suggested that a minimum of 80% be established as a reasonable level for the acceptable concentration of stannous ion in aged samples of this product line.

These guidelines were felt to be workable within the context of present production processes. Manufacturers were cooperative in supplying the requested information when available unless there was a potential concern regarding confidentiality and/ or pending patents. In such cases, verbal communication was established to help address these areas of review in at least a conceptual manner.

Some of our more recent findings are presented in Figure 10–5. Most products under review have largely been successful in improving the stability of their formulation; however, there appears to be an intrinsic instability associated with this product class, which is illustrated in the drift in stannous ion content with time for each product. There also appears to be some difference between products as to the rate of this change.

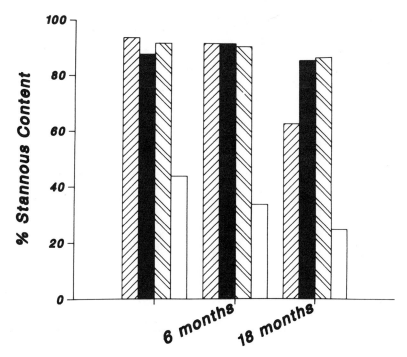

Fig. 10–5. Measured mean levels of free stannous ion in four commercial 0.4% stannous fluoride gels. Products were assayed for stability over 18 months.

APF GELS AND THEIR EFFECT ON PORCELAIN AND COMPOSITE RESTORATIONS

A variety of publications have reported that porcelain or composite exposed to acidulated phosphate fluoride results in increased surface roughness,[18,19] loss of esthetics,[20] and loss of weight[21] in the restorative materials. This is not surprising since the storage of fluoride in glass has been reported to result in the etching of the glass surface and loss in fluoride content.[22] APF gels contain hydrofluoric acid, which is known to etch glass, a major component of porcelain.

The Laboratories of the Council on Dental Therapeutics and the Council on Dental Materials, Instruments, and Equipment have initiated a study to examine and help quantify the degree of surface alteration. The following discussion relates to this work, which is still in progress.

Method. Porcelain specimens have been obtained from the five commercial manufacturers listed in Table 10–7. To date, two of these commercial porcelains have been examined. Both products have been treated with three fluoride gels: two APF gels (Luride and Gel II) and one neutral 2.0% NaF gel (Nupro-Neutral). The chemical composition of these treatments are presented in Table 10–8.

Five samples of each porcelain were evaluated for the degree of change in specular reflectance as a function of both time of treatment application as well as wavelength. Sample selection was based upon uniformity of surface structure and color as judged by visual observation. Samples were required to be smooth and flat. These samples were then slightly adapted to fit the sample holder as well as to permit replacement

Table 10–7. Porcelain Products Under Review

HOWMEDICA, INC.
   (Micro-Bond Natural Ceramic·
   Fired to a Natural Glaze)
JOHNSON & JOHNSON
   (Cermaco B Gingival 59)
NEY
   (Uncolored Porcelain Batch No. 157)
UNITEK
   (VMK68, Shade D$_2$, Lot #453)
JELENKO
   (Vita-Lumin Porcelain)

Table 10–8. Chemical Composition of Fluoride Treatments used in Porcelain Study

	PH	F⁻ Labeled	F⁻ Found	VISCOSITY 30 RPM	60 RPM
Luride	3.6	1.2%	100.3%	11,337	7,164
Gel II	3.8	1.23%	96.1%	7,611	4,681
Nupro-Neutral	7.0	0.9%	101.0%	8,595	5,348

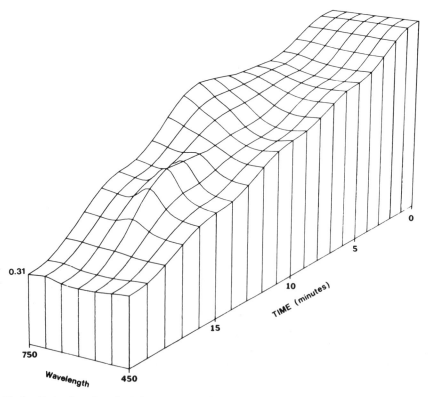

Fig. 10–6. Projection plot of relative specular reflectance presented as a function of treatment time and wavelength. Intensity changes relative to initial normalized reflectance at time equal to zero. Micro-Bond Natural Ceramic (Howmedica, Inc.) was treated with Gel II APF gel.

of samples in the sample holder. An extensive period of time was devoted to the minimization of variability in reflectance following sample replacement.

Samples were then washed with water, wiped with a Kimwipe and then dried with a heatgun for 2 min to make sure samples were free of surface moisture. They were then repositioned in the sample holder and wiped one last time.

Specular reflectance was monitored using a single beam Aminco Bowman spectrophotofluorometer in which the sample chamber was modified permitting the measurement of reflectance off a solid sample at an incident angle equal to the reflectance angle of 45°. In order to monitor the characteristic intensity and wavelength dependence of the light source and photomultiplier, both a control sample and an optical mirror were sandwiched between each observation. Because this study required quantified intensity measurements, this technique was implemented as an attempt to account for the modest variation of the light intensity from our high intensity xenon lamp with time.

Due to the fluorescent character of the samples, intensity measurements were made at 50 nm increments between 450 and 750 nm. The measurements were initiated at 450 nm due to the need to use an ultraviolet cut-off filter to minimize fluorescence in the visible range. This filter had an absorption band that substantially reduced transmission at 400 nm. Furthermore, sampling above 750 nm was not performed due to the poor signal-to-noise ratio, a result of the relatively weak output at these wavelengths from the xenon lamp.

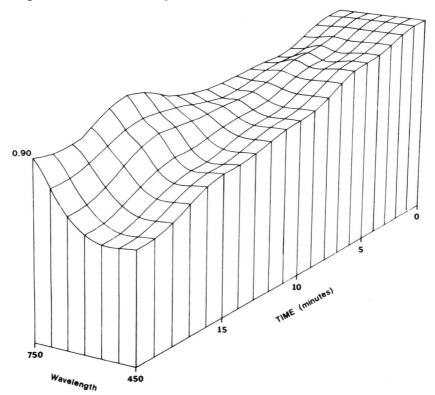

Fig. 10–7. Projection plot of relative specular reflectance presented as a function of treatment time and wavelength. Micro-Bond Natural Ceramic (Howmedica, Inc.) was treated with Luride APF gel.

The light source was fixed at each of these wavelengths and the reflectance intensity was then scanned over a 100 nm range centered about the wavelength of interest. In general, the emission spectra was slightly shifted and skewed towards higher wavelengths. Because the slope in change of intensity at the excitation wavelength was large, the maximum intensity reading rather than the intensity at the excitation wavelength was consistently utilized throughout in the analysis of this data.

Following reflectance measurements over the wavelength range of 450 to 750 nm, samples were then coated with one of the selected fluoride solutions for 4 min. The samples were then washed with water to remove all of the test product, wiped, and then dried for 2 min. Samples were then remounted in the sample holder and measured once again for specular reflectance. This procedure was repeated five times, yielding reflectance information at 4, 8, 12, and 20 min treatment times.

CONCLUDING REMARKS

Figures 10–6 to 10–9 illustrate the variation in intensity of specular reflectance for the porcelains treated with the two APF gels. The relative degree of reflectance at a wavelength of 450 nm of the Howmedica product as measured after 20 min of treatment with Gel II and Luride was found to be 30 and 59% of the initial reflectance level, respectively. The degree of reflectance at 450 nm of the Johnson & Johnson product following these same two APF treatments was found to be 52 and 41% of

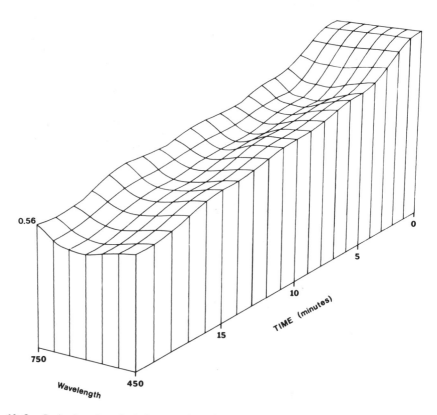

Fig. 10–8. Projection plot of relative specular reflectance presented as a function of treatment time and wavelength. Cermaco B Gingival 59 (Johnson & Johnson) was treated with Gel II APF gel.

initial intensity, respectively. Note further that with the exception of the Luride treatment on the Howmedica product, there appears to be a rather uniform decrease in intensity with time over the entire visible spectra. This specific case may be a result of changes at the microscopic level in the surface structure, which could be wavelength-dependent.

Changes in the relative degree of specular reflectance due to treatment of these porcelains with the near neutral fluoride product was found to be very small.

Although there was detected more than a 50% reduction in the specular reflectance of the two porcelain samples following up to a 20-min exposure with APF gels, the degree of visual change in the gloss of these treated samples is only slight to modest. Changes in the samples are only detectable upon close comparison and if viewed in such a manner that reflection from nonhomogeneous external lighting is optimal.

The clinical implications of these changes in surface structure are still unclear. Although detectable changes in porcelains are observable over time periods clinically relevant, these changes in and of themselves are most probably not significant factors in the management of the average patient. There are, however, a number of additional factors to be considered. Changes in surface structure may well influence the degree of later potential staining and/or plaque accumulation. These secondary effects may be much more pronounced than the primary effect of surface erosion.

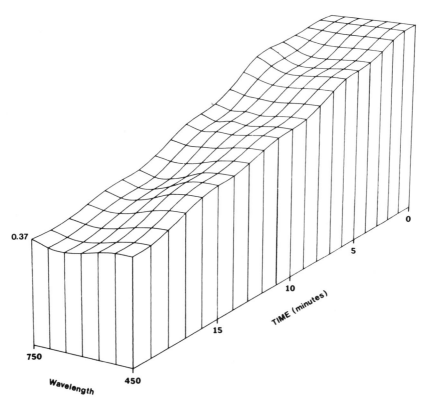

Fig. 10–9. Projection plot of relative specular reflectance presented as a function of treatment time and wavelength. Cermaco B Gingival 59 (Johnson & Johnson) was treated with Luride APF gel.

Additionally, patients under special treatment plans due to rampant caries may require special consideration. Extensive use of fluoride treatment (especially when low pH fluoride products are utilized) may result in substantially greater loss in the esthetics of the porcelain substrate. In such cases, additional precaution might be advisable to minimize the exposure of costly restorative work to acidulated fluoride products.

REFERENCES

1. Wellock, W.D., and Brudevold, V.: Caries increments, tooth discoloration, and state of oral hygiene in children given single annual applications of acid phosphate fluoride and stannous fluoride. Arch. Oral Biol., *10*:453–460, 1965.
2. Cartwright, H.L., Lindahl, R.L., and Bawden, J.W.: Clinical findings on the effectiveness of stannous fluoride and phosphate fluoride as a caries reducing agent in children. J. Dent. Child., *35*:36–40, 1968.
3. Horowitz, H.S., and Kau, M.C.W.: Retained anticaries protection from topically applied acidulated phosphate fluoride: 30- and 36-month post treatment effects. J. Prev. Dent., *1*:22–27, 1974.
4. Scott, D.B., Picard, R.G., and Wyckoff, R.W.G.: Studies of the action of sodium fluoride on human enamel by electron microscopy and electron diffraction. Public Health Rep., *65*:43, 1950.
5. Frazier, P.D., and Engin, D.W.: X-ray diffraction study of the reaction of acidulated fluoride with powdered enamel. J. Dent. Res., *45*:1145–1148, 1966.
6. Wei, S.H.Y., and Forbes, W.C.: X-ray diffraction analysis of the reactions between intact and powdered enamel and several fluoride solutions. J. Dent. Res., *47*:471–477, 1968.
7. Liang, Z., and Higuchi, W.I.: Kinetics and mechanism of the reaction between hydroxyapatite and fluoride in aqueous acidic media. J. Phys. Chem., *77*:1704–1710, 1973.
8. Loesche, W.J., Murray, R.J., and Mellberg, J.R.: The effect of topical fluoride on the percentage of *Streptococcus mutans* and *Streptococcus sanguis* in interproximal plaque samples. Caries Res., *7*:283–296, 1973.
9. Tinanoff, N., Brady, J.M., and Gross, A.: The effect of NaF and SnF_2 mouthrinses on bacterial colonization of tooth enamel: TEM and SEM studies. Caries Res., *10*:415–426, 1976.
10. Jenkins, G.N.: Theories on the mode of action of fluoride in reducing dental decay. J. Dent. Res., *42*:444–452, 1963.
11. Zahradnik, R.T., Propas, D., and Moreno, E.C.: Effect of fluoride topical solutions on enamel demineralization by lactate buffers and *Streptococcus mutans* in vitro. J. Dent. Res., *57*:940–946, 1979.
12. Stratemann, M.W., and Shannon, I.L.: Control of decalcification in orthodontic patients by daily self-administered application of a water-free 0.4% stannous fluoride gel. Am. J. Orthod., *66*:273–279, 1974.
13. Wescott, W.B., Starcke, E.N., and Shannon, I.L.: Chemical protection against post-irradiation dental caries. Oral Surg., *40*:709–719, 1975.
14. Zacherl, W.A., and McPhail, C.W.B.: Final report on the efficacy of a stannous fluoride-calcium pyrophosphate dentifrice. J. Can. Dent. Assoc., *36*:262–264, 1970.
15. Fogel, H.R., et al.: The relative caries-inhibiting effects of a stannous fluoride dentifrice on children: 2-year report of a supervised toothbrush study. J. Oral Ther. Pharmacol., *4*:175–180, 1967.
16. Hefferren, J.J.: Qualitative and quantitative tests for stannous fluoride. J. Pharm. Sci., *52*:1090–1096, 1963.
17. Shannon, I.L.: Water-free solutions of stannous fluoride and their incorporation into a gel for topical application. Caries Res., *3*:339–347, 1969.
18. Schlissel, E.R., Melnick, D.R., and Ripa, L.A.: In vitro effect of topical fluoride on porcelain surfaces. AADR Abs. #910, 1980.
19. Lacy, A., Copps, D., and Curtis, T.: Effects of topical fluorides on six low-fusing dental porcelains. AADR Abs. #602, 1982.
20. Gau, D.J., and Krause, E.A.: Etching effect of topical fluorides on dental porcelains: A preliminary study. J. Can. Dent. Assoc., *6*:410–415, 1973.
21. Kula, K., Nelson, S., and Thompson, V.: In vitro effect of APF gels on three composite resins. J. Dent. Res., *62*:846–849, 1983.
22. Hattab, F.: Stability of fluoride solutions in glass and plastic containers. Acta Pharm. Suec., *18*:249–253, 1981.

Chapter 11
FLUORIDES AND REMINERALIZATION

Leon M. Silverstone

Recent evidence supports the contention that one of the chief mechanisms by which fluoride reduces caries incidence is remineralization. Small enamel lesions that are either confined to the enamel or extend into the dentin in the interproximal regions cannot be detected by conventional clinical or radiographic techniques. Thus, such regions are diagnosed as sound tissue. Fluoride is concentrated into these demineralized areas which act as fluoride ion reservoirs and favor remineralization. Experimental studies have shown that although the presence of fluoride is important in enhancing remineralization, it is not necessary to employ high concentrations, because the frequent supply of low concentrations of fluoride appears to be optimal. The use of synthetic calcifying fluids on small lesions in vitro produces remineralization, which can be either surface limited or extend through the entire depth of a lesion. The calcium concentration of the calcifying fluid is important in determining which calcium phosphate phases are supersaturated, which in turn influences the degree of remineralization achieved. With high calcium concentrations, remineralization occurred in the surface layers of the lesion due to the rapid precipitation of precursor phases, which served to block surface pores, thereby preventing remineralization in depth. With low calcium levels, only the apatitic phases of the calcifying fluid were supersaturated and remineralization occurred in depth. The demonstration of crystals larger than those in sound enamel, produced as a result of remineralization, helps to explain the mechanism whereby remineralized lesions are more resistant to progression.

There is abundant evidence in the literature to show that small carious lesions can be arrested or reversed, a process referred to as remineralization.[1,2] Similarly, in vitro studies have shown that various types of experimental caries-like lesions can also be remineralized by exposure to either oral fluid or synthetic calcifying fluids.[3-7] One factor common to many of the results obtained on the remineralization of either enamel caries, artificial caries, or acid-etched enamel, is that the presence of fluoride ions greatly enhances the degree of remineralization achieved, and reduces the time period necessary for this mechanism to occur. The aim of this paper is to investigate the significance of fluoride in the remineralization of small caries-like lesions of human enamel in vitro.

Remineralization basically effects a small lesion by two processes. In the first process, the lesion is reduced in size by this phenomenon. In the second phase, the remineralized lesion becomes more resistant to progression. Thus, although the lesion is initially reduced in size, it is the latter mechanism that is probably most significant with respect to the clinical prevention of the lesion. If the lesion can be retained within the enamel, then it will not be diagnosed by clinical or radiographic means, and the clinician will be unaware of its presence.[8]

MATERIALS AND METHODS

Effect of Oral Fluid and Synthetic Calcifying Fluids on Lesion Remineralization In Vitro

Both human oral fluid and synthetic calcifying fluids were used in these studies. The oral fluid was obtained from an individual who was no longer caries-active and resided in a region having water fluoridation. Oral fluid is the term applied to the whole saliva sample obtained on expectoration, which contains additional components not found in the salivary excretion obtained directly from a gland.

The calcifying fluid was prepared from synthetic hydroxyapatite with a calcium/phosphate ratio of 1.63. Solutions having two different calcium concentrations were employed. These were 1.0 and 3.0 mM calcium. The phosphate concentrations remained at the fixed ratio of 1.63 (i.e., 0.61 and 1.84 mM, respectively). Sodium chloride was added to these fluids at a concentration of 200 mM; the pH was adjusted to 7.0 using 0.05 mM potassium hydroxide.

Fluoride ions were added to some of the samples of the synthetic calcifying fluids as well as to the oral fluid. These concentrations were either 0.05 mM (1 ppm), 0.5 mM (10 ppm), or 5 mM (100 ppm). Caries-like lesions were used in the experiments to provide a source of lesions having histological characteristics identical to those of enamel caries. Lesions were created in window regions on buccal and lingual surfaces of human teeth (Fig. 11–1, *1*). Teeth were exposed to gelatin gels acidified with lactic acid to pH 4.3 for exposure periods varying from 3 to 6 weeks (Fig. 11–1, 2) as described previously.[9]

After lesion production a single longitudinal section was removed from the center of the lesion to act as a control for the lesion (Fig. 11–1, *3*). Histological features of the lesions were recorded using the polarizing microscope, employing a range of imbibition media to explore the internal pore volume of the various zones. The cut faces on the remaining tooth halves were then varnished over so that just the outer surface of each lesion was exposed for experiment. Tooth halves were then exposed to either the oral fluid or the synthetic calcifying fluid (Fig. 11–1, *4*). In each experiment, one tooth half was exposed to the test fluid while the adjacent half of the same tooth was exposed to the identical fluid with added fluoride ions. Thus, for each test fluid, three groups of experiments were carried out, each comparing the test fluid with no added fluoride ions against the three concentrations of fluoride employed in this study. These tests were carried out with oral fluid as well as the two synthetic calcifying fluids. Exposure times of specimens to the various test fluids were either one, five, or ten consecutive 6-min increments as well as a single hour exposure. At the completion of a series of exposures, the enamel surfaces were brushed with water using an electric toothbrush and specimens were washed in agitated deionized water for 24 hours (Fig. 11–1, *5, 6*).

After the experiment, undemineralized longitudinal sections were obtained from the tooth halves (Fig. 11–1, *7*) using The Silverstone-Taylor Hard Tissue Microtome.[10] In this manner, it was possible to compare the control lesion with the effects of exposure to the test fluid both with and without the addition of fluoride ions. Quantitative imbibition studies were carried out on lesions using an Ehringhaus com-

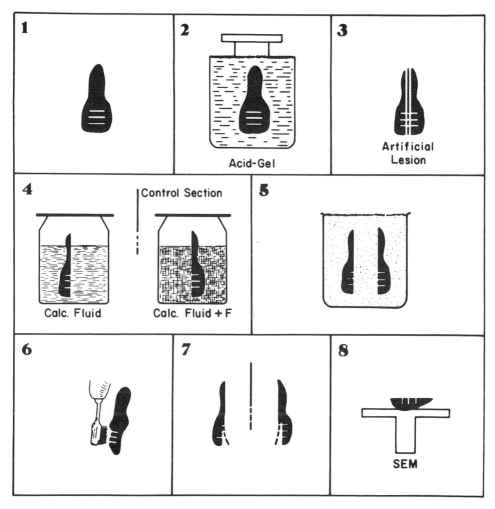

Fig. 11–1. Diagram of the experimental regimen used in the experiments comparing the effect of using calcifying fluids with or without fluoride ions.

pensator in conjunction with the polarizing microscope as described previously.[11] The histological zones of the lesions as well as contour maps were analyzed using quantitative image analysis techniques.

Test and control surfaces of lesions were also examined by scanning electron microscopy (Fig. 11–1, *8*) after vacuum coating with ultra-thin layers of platinum (~5 nm) using an ISI DS-130 SEM. In addition, material was prepared from longitudinal sections using the high resolution/microdissection technique described recently in order to examine crystals in situ within the zones of the lesion.[7,12]

Effect of Fluoride and Calcifying Fluids on Lesion Progression In Vitro

In this group of experiments, the "single-section" technique was used creating lesions in vitro on longitudinal sections rather than employing whole teeth.[13,14] In this manner, lesions can be created with relatively short exposure times measured in days instead of weeks. After lesion formation, the effect of test fluids on the surface above the lesion can be evaluated during a subsequent lesion progression cycle. Thus, it is possible to observe a series of progressions of a single lesion. Since the lesion acts as its own control, the test fluid can be evaluated with great accuracy.

The experimental regimen is shown in Figure 11–2. A series of longitudinal sections were prepared using The Silverstone-Taylor Hard Tissue Microtome (Fig.

LESION FORMATION
ON SINGLE SECTIONS

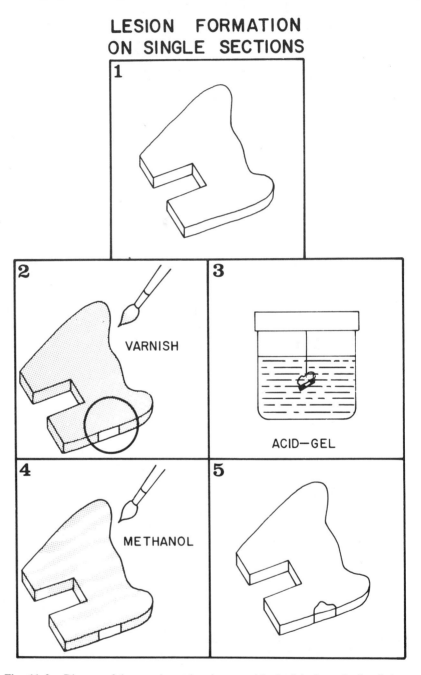

Fig. 11–2. Diagram of the experimental regimen used in the "single-section" technique.

11–3A). With this apparatus it is possible to prepare 15 or more undemineralized serial sections through a molar tooth (Fig. 11–3B). A single section (Fig. 11–2, *1*) is painted with an acid-protective varnish leaving just the outer surface of a test site exposed (Fig. 11–2, *2*). The section is then suspended in a 17% gelatin gel acidified

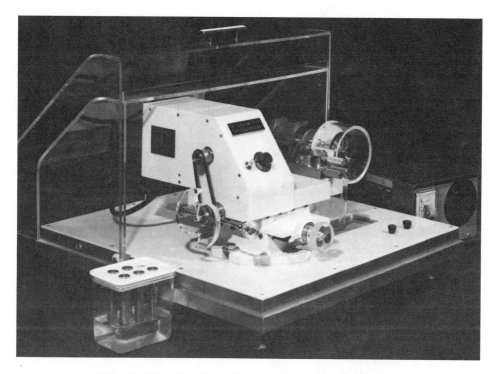

Fig. 11–3,A.   The Silverstone-Taylor Hard Tissue Microtome.

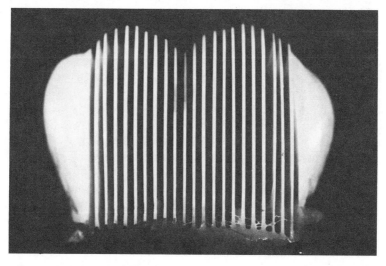

Fig. 11–3,B.   A molar tooth from which 21 serial undemineralized sections have been prepared using The Silverstone-Taylor Hard Tissue Microtome.

with lactic acid to pH 4.3 and containing 0.5 g/liter synthetic hydroxyapatite (Fig. 11–2, *3*). Sections were removed after a 3-day period; the varnish was removed in methanol so that lesion parameters could be recorded (Fig. 11–2, *4*). Sections were then revarnished, leaving the original test window exposed and then returned to the same acidified gel for 3 more days for lesion progression. After this, lesion parameters were again recorded. The surface superficial to the lesion was then exposed to one of three test regimens.

(1) A solution of 0.05 mM fluoride ions at pH 7.0 (1 ppm).

(2) A 3 mM calcifying fluid. This was prepared as described previously in this section and contained 0.05 mM fluoride ions.

(3) A 1 mM calcifying fluid, also prepared in the same manner as described previously and similarly containing 0.05 mM fluoride ions.

Sections were then once again exposed to the acidified gel artificial caries technique for lesion progression over a final 3 days exposure. Lesions were evaluated for the final time after this regimen.

Results

Experiments A. The experiments showed conclusively that small, intact surface lesions in human dental enamel were able to remineralize in vitro.

When oral fluid was used, results of remineralization were found to be restricted to the outer enamel surface. Results of the quantitative imbibition studies are shown in Plate I, A. The area of the body of the lesion has been plotted in green. Superimposed on this is the area of the body region after exposure to oral fluid, which has been reproduced in red. After exposure to oral fluid containing added fluoride ions, the body region has been superimposed on the diagram employing a blue contour. Plate I, A shows the final imbibition graph with all three regions included. The curve for the lesion after exposure to oral fluid, both with and without fluoride ions, differs from the control. This is indicated by the fact that neither the red nor the blue contours are completely superimposed over the control contour shown in green. When oral fluid alone was used, there was a reduction in area of the body of the lesion. There was little difference, however, between using oral fluid with or without added fluoride ions. This is shown by the fact that the red contour is not very large. When oral fluid alone was used, there was a reduction in area of the body of the lesion of 7.2% relative to the control. When fluoride ions were added to oral fluid, there was a greater degree of remineralization, as can be seen from the blue contour in Plate I, A. There was a 10.1% reduction in area of the body of the lesion. Thus, the addition of fluoride ions resulted in a further decrease in size of the superficial part of the body of the lesion. Mean results for a group of ten lesions in these experiments were 8 and 11%, respectively.

When the synthetic calcifying fluids were employed, a much greater degree of remineralization occurred. Plate I, B shows the effect of the 3 mM calcium calcifying fluid. When the calcifying fluid was used without the addition of fluoride ions, there was a significant reduction in the size of the body of the lesion. Plate I, B shows a typical example from these experiments employing the same color code as used with oral fluid. The control lesion has been depicted in green once again. Over this has been superimposed the effect of the calcifying fluid with no added fluoride, shown

Plate I

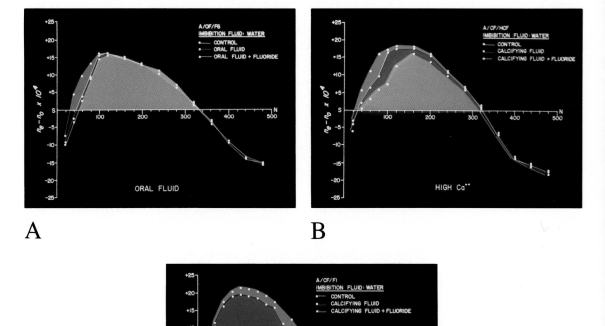

A   B

C

A. Imbibition graphs showing birefringence curves for an artificial lesion after exposure to oral fluid both with and without fluoride ions. The abscissa S-N represents a traverse from the enamel surface (S) through the lesion and into normal enamel (N), measured in micrometers. The ordinate shows observed birefringence; the higher the graph extends into the upper positive quadrant, the greater is the degree of demineralization. Results for the control lesion are shown by the green contour over which is superimposed the effect of exposure to oral fluid containing no fluoride, shown by the red contour. The effect of using oral fluid containing 100 ppm fluoride is shown by the blue contour. The red and blue contours are similar in size, the red contour being slightly larger. Thus, oral fluid containing fluoride at 100 ppm was slightly more effective than using oral fluid alone. The effect of both regimens has been limited to the surface layers of the lesion.

B. Similar graph constructed for a lesion exposed to the synthetic calcifying fluid containing 3 mM calcium. The effect of adding fluoride at a level of 1 ppm has more than doubled the degree of reduction in lesion porosity. In both cases, however, remineralization has been limited to the superficial aspect of the body of the lesion.

C. Similar graph constructed for a lesion exposed to the 1 mM calcium calcifying fluid. As with the previous two graphs, the control is shown in green, the effect of the calcifying fluid without fluoride ions in red, and the identical fluid with 1 ppm fluoride in blue. The addition of fluoride has produced a dramatic reduction in lesion size and porosity relative to using the calcifying fluid without fluoride. In both cases, however, remineralization has occurred throughout the entire depth of the lesion.

Plate I (continued)

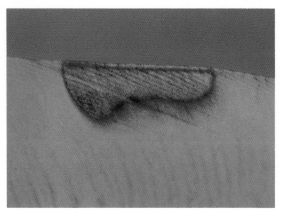

D

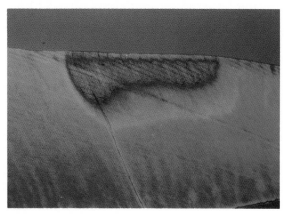

E

F

D. Longitudinal undemineralized section through a human tooth showing a caries-like lesion created through the surface of the section using the "single-section" technique. The section is examined in water with the polarizing microscope. The lesion was created in an acidified gel after an exposure period of 18 days. The lesion shows a negatively birefringent surface zone with a depth of 20 μm superficial to the positively birefringent body of the lesion.

E. Same section as in D after the outer surface of the test site was exposed to a synthetic calcifying fluid for five consecutive 6-min exposures. The section is examined under identical conditions to that in D. The body of the lesion has been reduced in area by 34% compared with the control state. The surface zone has also increased in depth from 20 to 35 μm and shows a reduction in porosity.

F. Same section as in the previous figure after the outer surface of the test site was exposed to the calcifying fluid for a further five 6-min increments. Most of the lesion now shows as a region of pseudo-isotropy. The body of the lesion has been reduced in area by 86% relative to the control seen in D. In addition, the surface zone shows a further increase in depth to 50 μm. This demonstrates the accuracy of the "single-section" technique for experiments on lesion initiation, progression, and remineralization.

by the red contour. In this example, there was a reduction in area of the body of the lesion of 12%. As with the experiments using oral fluid, the changes have occurred at the surface of the lesion. When the same calcifying fluid was used containing 0.05 mM fluoride, a further reduction was found, as shown by the blue contour. In this example, there has been a 34% reduction relative to the control. In spite of this significant increase in remineralization, changes are still limited to the surface region of the lesion. Mean reductions found for a group of ten lesions in this part of the study were 9 and 24%, respectively.

Plate I, C shows the effect of employing the 1 mM calcium calcifying fluid in these experiments. The same color-coded contours have been used. When the calcifying fluid was used without the addition of fluoride ions, there was a 20% reduction in area of the body of the lesion relative to the control. On examining the red contour relative to the control (shown in green), however, it is obvious that changes have occurred at the advancing front of the lesion instead of at the enamel surface. This is in striking contrast to the previous two groups. When fluoride ions were added, the reduction in the body of the lesion was dramatic, being 86% relative to the control. Similarly, the reduction in size of the body of the lesion has occurred throughout the entire depth of the lesion and not just at the surface. Mean reductions found for a randomly selected group of ten lesions were 22 and 72%, respectively. The results found with both oral fluid and the synthetic calcifying fluids are listed in Table 11–1.

With the synthetic calcifying fluids, the greater effect achieved by the addition of fluoride ions occurred irrespective of the level of fluoride added. No greater degree of remineralization occurred by increasing the fluoride level within the calcifying fluid. With oral fluid, the only experiments in which there was a significant increase in remineralization were those in which fluoride was added at a level of 100 ppm. There were no significant differences when using oral fluid alone relative to oral fluid containing fluoride added at either 1 or 10 ppm.

When oral fluid was used in these experiments, a firmly adherent layer was formed on the enamel surface after a single 6-min exposure period. This layer became thicker and more irregular with increasing exposure time. Maximum changes in the lesions occurred after five 6-min exposure increments. When the synthetic calcifying fluids were used, maximum changes were recorded after the second series of five 6-min exposure increments. Approximately 75% of this degree of change, however, had occurred after the first five exposure increments. When a lesion was exposed for a single 1-hour increment, the changes produced exceeded a single 6-min exposure, but were significantly less than was produced after the first series of five 6-min exposure increments.

Table 11–1. Mean Reductions in Area of the Body of the Lesion for Groups of Lesions Selected at Random

Calcifying fluid	% Reduction of Body
Oral Fluid	8
Oral Fluid + F	11
3 mM Ca	9
3 mM Ca + F	24
1 mM Ca	22
1 mM Ca + F	72

Examination of lesions by scanning electron microscopy showed some interesting findings. After exposure of enamel surfaces to human oral fluid, a surface coating was readily visible on the enamel and remained even after brushing with an electric toothbrush and after a 24-hour wash period. Figure 11–4 is a scanning electron micrograph of the enamel surface after exposure to oral fluid. The entire enamel surface was covered with an irregular organic coating appearing as a series of ridges and valleys. Examination at a higher magnification showed the dense network of material (Fig. 11–5). In other regions, a series of small rounded and elongated bodies were seen scattered in intervals over the surface coating. When examined in more detail, the "rounded" bodies were seen to have straight sides, and were usually hexagonal in outline (Fig. 11–6). Some of these crystals had four to six sides but the majority were eight-sided. Adjacent to these hexagonal crystalline deposits were elongated bodies. The hexagonal deposits varied from 150 to 300 nm in diameter, whereas the elongated bodies were 100 to 200 nm in diameter.

When specimens were exposed to the synthetic calcifying fluids, such surface coatings were not detected. When the 3 mM calcium calcifying fluid was employed with the addition of fluoride, however, crystalline material was observed on test enamel surfaces. Figure 11–7 is a scanning electron micrograph of the crystalline deposits found under these conditions. Clusters of plate-like material were seen contiguous

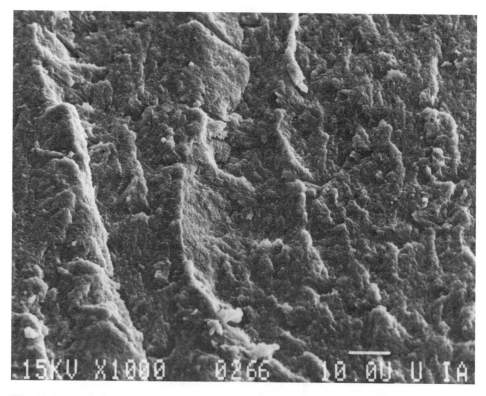

Fig. 11–4. Scanning electron micrograph showing the enamel surface superficial to a lesion after five consecutive exposure increments to oral fluid. The surface is coated with an irregular organic coating which appears as a series of ridges. Space bar denotes 10 μm.

Fig. 11–5. Higher power scanning electron micrograph showing part of the organic coating after exposure to oral fluid. The coating appears as a dense irregular mass of organic material. Space bar denotes 1 μm.

with and growing from the enamel surface. They were also seen to vary from small rosettes to large plates. This crystalline deposit was not seen with the specimens exposed to the 1 mM calcium calcifying fluid.

Figure 11–8 is a scanning electron micrograph from a sliver of material prepared from a lesion remineralized with the 1 mM calcium calcifying fluid containing fluoride ions. This was prepared using the high resolution/microdissection technique previously described.[12] A single enamel prism can be seen in which the constituent crystals can be detected with ease. When these were examined at high magnification, it was obvious that there was a significant increase in crystal diameters throughout the lesion (Fig. 11–9).

Experiments B. The series of experiments carried out using the "single-section" technique showed the effects of the various test regimens on lesion progression. When the 0.5 mM fluoride solution was used between the first lesion progression cycle and the second lesion progression cycle, there appeared to be relatively little effect in terms of modifying the lesion progression rate. The mean reduction in lesion progression rate for a group of ten lesions was only 2%. In contrast, the 3 mM calcium calcifying fluid containing 0.05 mM fluoride had a significant effect in retarding lesion progression. In a series of ten lesions examined over two progression cycles per lesion, the progression rate was reduced 23%. When the 1 mM calcium

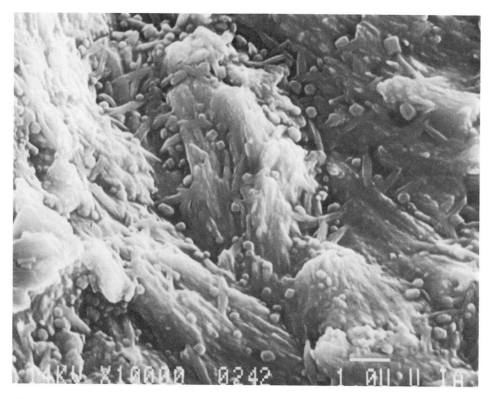

Fig. 11–6. Scanning electron micrograph of part of the coating produced on the enamel surface by oral fluid. Numerous elongated and rounded bodies can be seen. The rounded bodies appear to be hexagonal in outline in many cases, whereas others have four to six sides. These precipitates were seen in various regions of the coating and were positioned superficially in the more dense and structureless organic material. Space bar denotes 1 μm.

calcifying fluid was used, also containing 0.05 mM fluoride, the lesion progression rate was reduced 73%. These results are shown in the histogram in Figure 11–10.

The "single-section" technique is an ideal method to accurately document lesion modification since the lesion acts as its own control. Plate I, D shows a caries-like lesion, which was created through the outer surface of a single section after an 18-day exposure. This lesion is examined in water with the polarizing microscope.

The body of the lesion is seen as the region exhibiting positive birefringence when the section is examined in water with the polarizing microscope (Plate I, D). This region has a minimum pore volume of 5% at its periphery, increasing to well in excess of 25% towards its center. After the first series of exposures to the calcifying fluid, the body of the lesion showed a significant reduction in size with 34% of the original region displaying a pore volume of 5% or less. Since the body region is, by definition, the region of positive birefringence seen in water, this demonstrates that the body region was reduced to 66% of its original size. After the second series of five 6-min exposure increments to the calcifying fluid, there was a further reduction in both size and porosity of this region. The body of the lesion (Plate I, F) now appears almost exclusively as a region of pseudoisotropy with only 14% of its original area displaying positive birefringence. Thus, 86% of the original body of

Fig. 11–7. Scanning electron micrograph showing the enamel surface superficial to a lesion exposed to the 3 mM calcium calcifying fluid containing fluoride ions. Numerous crystals can be seen growing from the enamel surface. They vary from large plates to groups of rosettes. The space bar denotes 10 μm.

the lesion now exhibits a pore volume of 5% or less. Therefore, over the period of ten 6-min exposures to the calcifying fluid the body of the lesion was reduced in area by 86%.

Over the series of these three photomicrographs shown in Plate I, D–F, there has also been a significant increase in depth of the surface zone. The surface zone is 20 μm in depth on the control lesion shown in Plate I, D. After the first series of exposures to the calcifying fluid (Plate I, E), the surface zone increased in depth to 35 μm as well as showing a reduction in porosity. After the second and final series of exposures to the calcifying fluid (Plate I, F), the surface zone showed a further increase in depth to 50 μm.

Discussion

When oral fluid was used as the calcifying fluid, small reductions in lesion size were found; however, these were limited to the surface of the lesion. The addition of fluoride ions to the oral fluid at concentrations of either 1 or 10 ppm produced no noticeable increase in remineralization. This may have been because the oral fluid sample contained fluoride, since it was obtained from a person residing in a fluoridated region. When the fluoride level was raised to 100 ppm, increases in remineralization were found and the superficial aspect of the body of the lesion was reduced

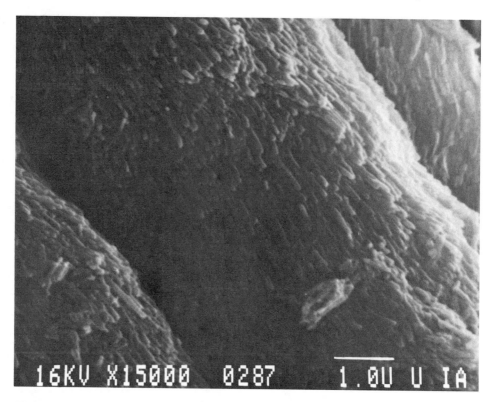

Fig. 11–8. Scanning electron micrograph showing a single prism from the region of the dark zone at the advancing front of a lesion. The crystals can be seen quite clearly. Space bar denotes 1 μm.

from 8 to 11%. A tenacious deposit was produced on the enamel surface which became thicker and more irregular with increased exposure time. The presence of this layer was probably why remineralization was restricted to the surface of the lesion and no further effects were found after the first series of five exposure increments. The bulk of the coating was organic in nature but crystalline deposits were seen scattered over the surface. These crystalline deposits occurred as either elongated or rounded bodies. The diameter of the crystal masses was larger than those of normal enamel crystals, but significantly smaller than micro-organisms. On closer examination, the rounded bodies were seen to have straight sides and were usually hexagonal in outline. These crystalline bodies were precipitated from the oral fluid, which is supersaturated with respect to both calcium and phosphate. The fact that they were not all hexagonal in outline, and comparable to hydroxyapatite crystals, is probably related to the presence of various calcification inhibitors present in oral fluid which had the effect of "poisoning" crystal surfaces, thus preventing the more normal types of crystal growth. It has been previously shown that body fluids containing physiological concentrations of either magnesium, zinc, bicarbonate, chromate, or pyrophosphate ions significantly slow the rate of remineralization.[15]

When the synthetic calcifying fluids were employed, a much greater degree of remineralization occurred. With the 3 mM calcium calcifying fluid, a crystalline deposit was seen on enamel surfaces. It has been shown by Silverstone and Wefel,[16] that with this calcifying fluid several of the more acidic calcium phosphate phases

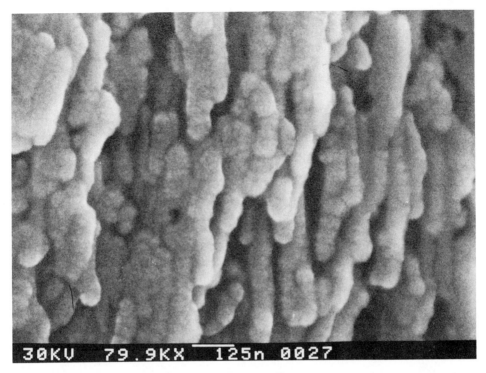

Fig. 11–9. Scanning electron micrograph showing a region of the body of the lesion from a specimen after exposure to the low calcium calcifying fluid containing fluoride ions. Groups of crystals can be seen, all of which are approximately twice the diameter of crystals in sound enamel. Multiple centers of recalcification can be seen along the crystal surfaces due to remineralization phenomenon. The space bar denotes 125 nm.

are supersaturated in addition to the apatite phases. Thus, acidic calcium phosphate phases and phase transformations precipitate on the enamel surface and tend to block surface pores. This is the reason why remineralization of lesions was limited to a depth of approximately 100 μm. In time, these precursors undergo phase transformation to apatite. Mean reductions for ten lesions selected at random were found to be 9% when the calcifying fluid was used without the addition of fluoride ions. When fluoride was added, the body of the lesion showed a 24% reduction in area.

With the 1 mM calcium calcifying fluid, only the apatite phases are supersaturated[16]; therefore, no precursor crystalline deposits were found on enamel surfaces. Thus, remineralization was seen to occur throughout the entire depth of a lesion and this synthetic fluid was able to produce more favorable results. Mean reductions were found to be 22% and, when fluoride was added, this increased to 72%. Thus, the addition of fluoride ions to the synthetic calcifying fluids had a significant effect in promoting remineralization.

Of interest is the fact that the greater effect achieved by the addition of fluoride ions occurred irrespective of the level of fluoride added. Thus, similar results occurred irrespective of whether fluoride was added at 1, 10, or 100 ppm. No greater degree of remineralization occurred by increasing the fluoride level within the calcifying fluid. In fact, current work employing higher concentrations of fluoride in the calcifying fluids has shown a definite reduction in the degree of remineralization

REDUCTION IN LESION PROGRESSION
RATE IN VITRO

EFFECT OF CALCIFYING FLUIDS
CONTAINING FLUORIDE, AND
FLUORIDE ALONE

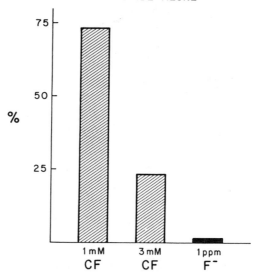

Fig. 11–10. Histogram to show the reduction in lesion progression rate in vitro after the use of synthetic calcifying fluids containing fluoride and the use of fluoride alone.

achieved relative to using lower levels as employed in these experiments.

Studies employing the "single-section" technique have proved to be extremely useful. This is the only technique in which it is possible to examine the sound enamel prior to creating a lesion over the identical test site. After lesion initiation, the very same lesion can be advanced over a series of progressions. At any specific point in time, test agents such as calcifying fluids can be introduced to the surface superficial to the lesion. Thus, a further bout of lesion progression will evaluate the effect of the test agent.

An example of such a lesion produced on a single longitudinal section is shown in Plate I. The lesion shown (Plate I, A) exhibits the two histological zones of enamel caries, which can be observed after imbibition with water, namely the surface zone and the body of the lesion. The surface zone has been defined by Silverstone as a region of negative birefringence when the section is examined in water.[17,18] The fact that the surface layer of a lesion appears as a radiopaque zone on a microradiograph is by no means indicative of the presence of a surface zone comparable to that appearing in the lesion of enamel caries, as will be discussed later.

When this section (Plate I, A) was examined in quinoline, both a translucent zone and a dark zone were observed at the advancing front of the lesion. Therefore, this lesion appeared indistinguishable from enamel caries based on the appearance of the four classical histological zones of the lesion.[9] This is the first report of the creation of a lesion on a single section, which appears histologically indistinguishable from enamel caries.

Plates I, E and F demonstrate the significant reduction in size and porosity of the lesion by exposure to a 1 mM calcium calcifying fluid containing 0.05 mM fluoride ions. Over a period of ten 6-min exposure increments the most heavily demineralized region of the lesion, namely the body of the lesion, was reduced in area by 86%. At the same time, the surface zone of the lesion increased in depth from 20 to 50 μm, and showed a reduction in porosity to less than 1% pore volume.

The surface zone of a lesion is probably the most important region in determining the fate of the lesion, since it acts as a rate restricting membrane. Even in a small, radiographically undetectable lesion the subsurface body of the lesion approaches a pore volume of about 25%. The surface zone in such a lesion is negatively bire-fringent when examined in water (i.e., it has a pore volume of less than 5%). Often the surface zone is close to 1% pore volume. Therefore, if remineralization is to occur, ions must pass through this well-mineralized barrier. Thus, it is important to simulate this pore volume distribution in the lesion when using artificial lesions in such experiments. If lesions are produced having heavily demineralized surface layers, a feature which is not found in enamel caries, it may well be a simple process to introduce ions through the surface layer and modify the underlying body of the lesion.

Figure 11–11 is a diagram of a lesion seen by two techniques, polarized light microscopy and microradiography. The histological features of the lesion show the presence of a translucent zone and a dark zone at the advancing front of the lesion. Superficial to this is the body of the lesion. This diagram shows a point in time when

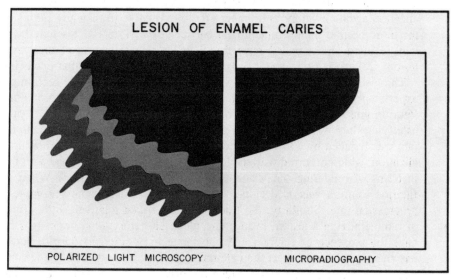

Fig. 11–11. Diagram to demonstrate the appearance of an enamel lesion viewed by two different techniques. Using polarized light microscopy (left-hand side), the advancing front of the lesion shows both a translucent zone and a dark zone, neither of which can be detected on a microradiograph (right-hand side). The lesion indicated in the diagram is one that had had heavy surface demineralization; as a result, there is no longer a surface zone. There is no indication of the loss of the surface zone, however, when the lesion is viewed by microradiography. Thus, the microradiograph will always detect the presence of the body of the lesion. The radiopaque surface layer may, or may not, coincide with the surface zone of the lesion. Therefore, microradiography will detect one, and possibly two of the four histological zones of a small enamel lesion. The presence of a radiopaque surface layer on a microradiograph is not an accurate criterion of whether the lesion has a surface zone.

the surface layer has been so heavily demineralized that there is no surface zone in the lesion. This usually occurs just prior to cavitation. The right-hand side of the diagram shows how a microradiograph of this lesion would appear. The body of the lesion appears as a subsurface region of radiolucency while the surface of the lesion is radiopaque. This would occur provided that there was a mineral content differential between the surface layer and the subsurface region, which is always the case. Thus, if the body of the lesion had a mineral loss of 70%, and the surface layer exhibited 60% mineral loss, then the microradiograph would still show a radiopaque surface layer. Thus, it is impossible to determine if a lesion has a surface zone comparable to that in the carious lesion from microradiographic evidence alone. This is why microradiographic examination alone of in vitro lesions is a poor criterion for determining their similarity to the lesion of enamel caries.

Figure 11–12A shows a section through an enamel lesion produced in a buffered acid system, which bears only slight resemblance to a carious lesion. The section is examined in water and has no surface zone since the body of the lesion extends right up to the outer enamel surface. In addition, there is evidence of surface loss. There was no evidence of a dark zone when the lesion was examined in quinoline. Thus, this lesion is a poor facsimile of a lesion of enamel caries. Figure 11–12B is a microradiograph of the lesion, showing the presence of a radiopaque surface layer superficial to the body of the lesion. Thus, from the microradiographic evidence alone it might have been concluded that this lesion had a surface zone similar to that occurring in the small enamel lesion. This point is not just of academic interest. If experimental lesions have extremely porous surface layers, the effect of test remineralization agents is likely to be more effective relative to their use on lesions having the more normal highly mineralized barrier at the surface. This may be the reason why Gelhard and Arends[19] have found that remineralization of their experimental lesions in vitro is of a greater order of magnitude relative to that occurring in vivo.

The histogram in Figure 11–10 shows the results of the single-section progression experiments. When the 3 mM calcium calcifying fluid was used, containing 1 ppm fluoride ions, lesion progression rate was reduced 23%. When the 1 mM calcium calcifying fluid was used containing the same fluoride concentration, the progression rate was reduced by 73%. It would appear that the deeper, more thorough remineralization, which occurred with the low calcium calcifying fluid, had a profound effect in terms of rendering the lesion more resistant to progression. When the 1 ppm fluoride solution was used alone, however, there was virtually no effect on lesion progression rate. Similarly, the fluoride solution had relatively little effect in terms of remineralizing a lesion, even under ideal laboratory conditions. It is essential to have the presence of calcium and phosphate in addition to fluoride ions. Brown[20] has shown that lowering of the calcium hydroxide activity within a lesion would be beneficial because this would reduce the driving force for diffusion of calcium out of the lesion and of protons into the lesion. The presence of fluoride ions tends to both lower the pH and the calcium hydroxide activity and to raise the phosphoric acid activity. When applied to the solution phase within a lesion, the ultimate effect is to alter the rates of diffusion in a way that diminishes the rate of lesion formation or increases the rate of remineralization. Thus, caries preventive agents used for remineralization are not likely to be highly effective if they rely solely on the presence of the fluoride ion.

What is the reason for the greater resistance to progression of a remineralized lesion? It was initially believed that the increased fluoride level in the lesion was responsible for this. Recent studies using lesion microdissection techniques coupled with high resolution scanning electron microscopy have provided relevant information.

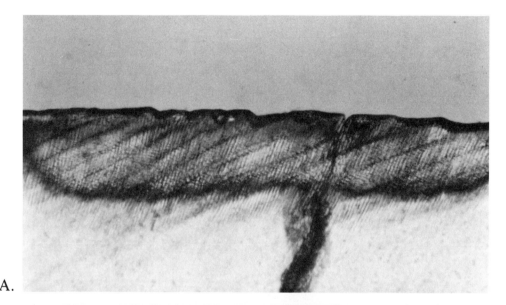

A.

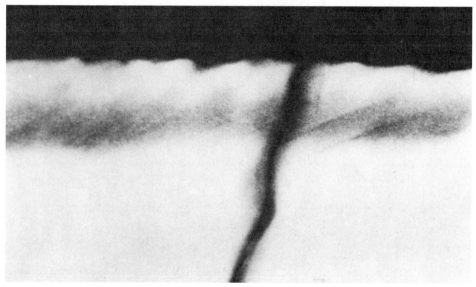

B.

Fig. 11–12. A. A longitudinal section of an artificial lesion produced in buffered lactic acid examined in water with the polarizing microscope (× 125). The body of the lesion appears positively birefringent, and this region spreads superficially to include the surface layer. Thus, there is no surface zone in this lesion. B. Micrograph of the same section. The subsurface region of the lesion appears as an area of radiolucency in contrast to the surface layer, which appears radiopaque.

During lesion formation the submicroscopic crystals are affected by acid dissolution, resulting in a diminution or thinning along their C-axes. This results in smaller, narrower crystals, which in turn are responsible for the increased pore volume in the lesion with increasing demineralization. Using a microdissection technique it has been possible to prepare specimens from sections through lesions in such a manner that crystals within the zones of the lesion could be examined directly by scanning electron microscopy.[12] Silverstone[12] found that crystals within both the dark zone (Zone 2) and the surface zone (Zone 4) of the carious lesion in human enamel

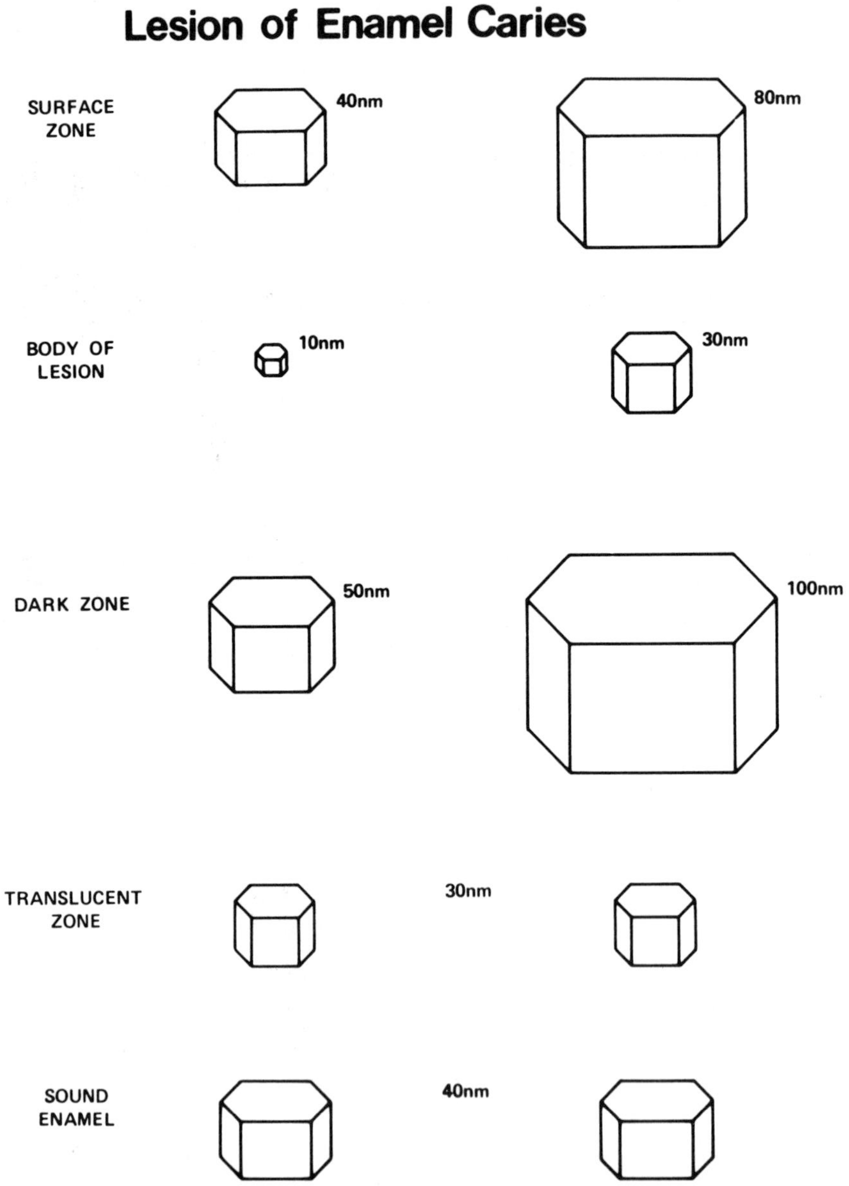

Lesion of Enamel Caries

Fig. 11–13. Diagram to illustrate the relative crystal diameters in sound enamel and in the four histological zones of the enamel lesion.

were significantly larger in diameter than either those in the rest of the lesion, or in the sound enamel (Fig. 11–13). The relevant point is that the above two zones have previously been reported as being formed as a result of remineralization phenomena.[2] When test halves of lesions were exposed to a synthetic calcifying fluid, there was a significant increase in crystal diameters throughout the entire lesion (Fig. 11–14).

Figure 11–9 shows a high power scanning electron micrograph of crystals within the body of the lesion after exposure to the 1 mM calcium calcifying-fluid-containing fluoride. These crystals have an average diameter of approximately 70 nm, even

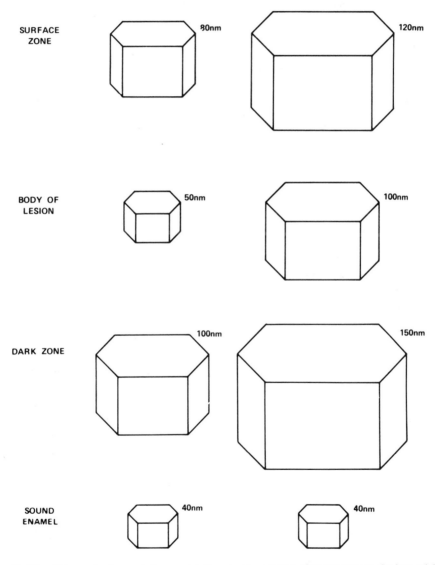

Fig. 11–14. Diagram to show relative crystal diameters in a lesion after exposure to the low calcium calcifying-fluid-containing fluoride ions.

though this was previously the most heavily demineralized region of the lesion, having crystal diameters of less than 40 nm as found in sound enamel. It is also possible to see separate recalcification foci along the crystals. There is evidence that octacalcium phosphate is involved in the growth of enamel crystals.[21] It has been suggested[22] that one role of fluoride ions is to eliminate the presence of octacalcium phosphate by facilitating its conversion to apatite. The conversion of this relatively soluble phase (OCP) to apatite would help to account for the fact that the presence of only a few fluoride ions per 100 unit cells has a profound effect on caries incidence.

Since the presence of large crystals, relative to small crystals, produces a favorable change in the surface area to volume ratio, and as these new crystals are almost certainly fluoridated hydroxyapatite, these factors must play a major role in increasing the resistance of the lesion to progression.

CONCLUDING REMARKS

Recent evidence supports the contention that one of the chief mechanisms by which fluoride reduces caries incidence is remineralization. Small enamel lesions confined to the enamel in the interproximal regions cannot be detected by conventional clinical or radiographic techniques.[8]

Figure 11–15 diagrammatically shows the size of a carious lesion when viewed by different techniques. In the upper three diagrams, the lesion is seen to penetrate about two-thirds of the depth of the enamel when examined histologically. When the same longitudinal section is examined by microradiography, the lesion appears smaller since only the body of the lesion is detected by this technique. Thus, the advancing front of the lesion seen on a microradiograph coincides with the deep edge of the body region. Histologically, both the dark zone and the translucent zone are seen deep to the body of the lesion. The third diagram shows the results of bite-wing radiography. With a lesion of this size, no radiolucency will be detected on the radiograph and such a surface will be diagnosed as caries-free. The lower series of the three diagrams shows a more extensive lesion. Histologically, the lesion has penetrated into the dentin. Microradiography shows the lesion to be limited to the enamel, whereas bite-wing radiography indicates that the lesion is confined to the outer half of the enamel. Thus, regions of enamel with such lesions are diagnosed as sound tissue. It is into these demineralized areas that fluoride is concentrated, whereby these areas act as fluoride ion reservoirs and favor remineralization. Therefore, in a "caries-free" patient, many interproximal enamel regions are likely to have small lesions that are maintained at their histological subclinical size by continued remineralization. If fluoride is not available, lesions can progress in size to become detectable either clinically or by radiography. Once lesions are diagnosed, preventive techniques employing fluoride should be instituted, so as to favor remineralization. Lesions can then regress so as to "disappear" from radiographic detection.

Although the use of fluorides is important in enhancing remineralization, it does not appear to be necessary to use high concentrations. Recent clinical studies have supported the frequent supply of low concentrations of fluoride rather than the infrequent use of high concentrations. The results of this study support this contention, since the maximum degree of remineralization occurred with a low fluoride concentration of 0.05 mM (1 ppm) available at the enamel surface. Increasing the fluoride

Enamel Lesion Size Seen by Different Techniques

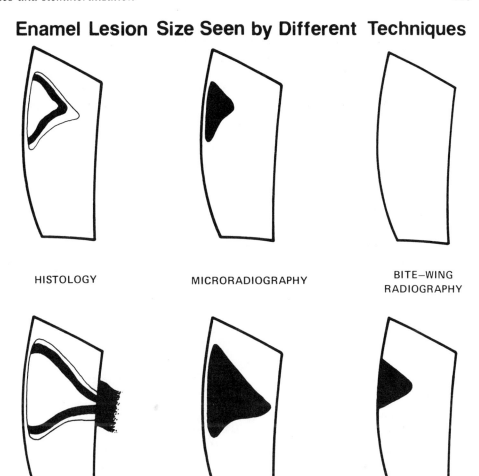

HISTOLOGY   MICRORADIOGRAPHY   BITE—WING RADIOGRAPHY

Fig. 11–15. Diagram to demonstrate the size of an enamel lesion as seen by different techniques. The most sensitive technique is examination by optical methods, preferably polarized light micropscopy, using quinoline as a mounting medium. In this manner both the translucent zone (Zone 1) and the dark zone (Zone 2) can be seen positioned in advance of the body of the lesion. The next sensitive technique is the use of microradiography, which demonstrates the depth of the body of the lesion. This is comparable to examining the lesion in water and viewing the section by optical methods. The least sensitive technique is the use of bitewing radiography. In the upper lesion, there would be no evidence of a lesion using bitewing radiography. In the lower, more extensive lesion, the bitewing radiographic appearance indicates a lesion positioned in the outer half of the enamel, whereas in fact the lesion has penetrated into the dentin.

level 100 times had no effect on the degree of remineralization achieved. Since it can take 3 or 4 years for a smooth surface lesion to progress to the stage where the dentin is invaded, there is adequate time to intercept the carious process. In this respect, the widespread use of fluoridated dentifrices are likely to have had a highly significant effect in reducing caries experience. The recent findings of lower caries incidence in many western nations have also been accomplished in regions where water fluoridation was either minimal or totally lacking. In such areas, fluoridated dentifrices must have been a major contribution in the frequent supply of low fluoride levels so as to enhance remineralization of lesions.

The demonstration of the presence of crystals in two of the four zones of the enamel lesion having diameters significantly larger than those of the sound enamel is of importance with respect to understanding why remineralized lesions become more resistant to progression. Thus, it appears that the initiation of a small subclinical lesion in the outer enamel paradoxically aids in preventing its progression. First, it acts preferentially, taking up significantly more fluoride than adjacent sound enamel. Therefore, the lesion acts as a fluoride ion reservoir and, during dissolution when calcium and phosphate ions are released, the presence of fluoride ions within the lesion favors remineralization. Secondly, if space were not available in the densely mineralized tissue, it would not be possible for crystal growth to occur. The fact that the crystal content of the lesion can be further modified, such that crystals within the body of the lesion can increase in size significantly, demonstrates the potential for increased remineralization. Thus, agents not currently available for clinical use could have a much greater potential in arresting and reversing the small subclinical lesion and, at the same time, increasing its resistance to further attack.

ACKNOWLEDGEMENT

This work was supported by NIDR Grant No. 7 R01 DE 06564.

REFERENCES

1. Backer-Dirks, O.: Posteruptive changes in dental enamel. J. Dent. Res., *45*:503–511, 1966.
2. Silverstone, L.M.: Remineralization phenomena. Caries Res., *11* (Suppl. 1):59–84, 1977.
3. Koulourides, T., Feagin, F., and Pigman, W.: Effect of pH ionic strength and cupric ions on the rehardening rate of buffer-softened human enamel. Arch. Oral Biol., *13*:335–341, 1968.
4. ten Cate, J.M., and Arends, J.: Remineralization of artificial enamel lesion in vitro. Caries Res., *11*:277–281, 1977.
5. Featherstone, J.D.B., Rodgers, B.E., and Smith, M.W.: Physicochemical requirements for rapid remineralization of early carious lesions. Caries Res., *15*:221–235, 1981.
6. Silverstone, L.M.: The effect of oral fluid and synthetic calcifying fluids in vitro on the remineralization of enamel lesions. Clin. Prev. Dent., *4*:13–22, 1982.
7. Silverstone, L.M.: Remineralization and enamel caries: New Concepts. Dental Update, *10*:261–273, 1983.
8. Silverstone, L.M.: The relationship between the macroscopic, histological, and radiographic appearance of interproximal lesions in human teeth: An in vitro study using an artificial caries technique. Ped. Dent., *3*:414–422, 1982.
9. Silverstone, L.M.: The structure of carious enamel, including the early lesion. *In* Oral Sciences Reviews, NO. 3. Dental Enamel. Edited by A.H. Melcher, and G.A. Zarb,. Munksgaard, Copenhagen, 1973.
10. Silverstone, L.M., and Taylor, R.: Preparation of thin, undemineralized, unembedded sections of human enamel: The Silverstone-Taylor Hard Tissue Microtome. J. Dent. Res., *60*:2, 1981.
11. Silverstone, L.M., et al.: Remineralization of natural and artificial lesions in human dental enamel in vitro: The effect of calcium concentration of the calcifying fluid. Caries Res., *15*:138–157, 1981.
12. Silverstone, L.M.: Remineralization and Enamel Caries: Significance of fluoride and effect on crystal diameters. *In* Demineralization and Remineralization of the Teeth. Edited by S.A. Leach, and W.M. Edgar. Oxford, England, IRL Press Ltd., 1983.
13. Featherstone, M.J., and Silverstone, L.M.: Creation of caries-like lesions in section of teeth using acid-gels. J. Dent. Res., *61*:279, 1982.
14. Featherstone, M.J., Silverstone, L.M., and Taylor, R.E.: Remineralization of caries-like lesions in vitro using the "single-section" technique. J. Dent. Res., *62* (Abst.):188, 1983.
15. Feagin, F.F., Walker, A.A., and Pigman, W.: Evaluation of the calcifying characteristics of biological fluids and inhibitors of calcification. Calcif. Tissue Res., *4*:231–244, 1969.

16. Silverstone, L.M., and Wefel, J.S.: The effect of remineralization on artificial caries-like lesions and their crystal content. J. Cryst. Growth., *53*:148–159, 1981.
17. Silverstone, L.M.: Surface phenomena in dental caries. Nature (Lond.), *214*:203–204, 1967.
18. Silverstone, L.M.: The surface zone in caries and in caries-like lesions produced in vitro. Br. Dent. J., *125*:145–157, 1968.
19. Gelhard, T.B.F.M., and Arends, J.: In vivo remineralization of artificial subsurface lesions in human enamel. J. Biol. Buccale, *12*:49–57, 1984.
20. Brown, W.E.: Physicochemical mechanisms in dental caries. J. Dent. Res., *53*:204–225, 1974.
21. Brown, W.E., Smith, J.P., Lehr, J.R., and Frazier, A.W.: Crystallographic and chemical relations between octacalcium phosphate and hydroxyapatite. Nature (Lond.), *196*:1050–1055, 1962.
22. Brown, W.E.: Crystal growth of bone mineral. Clin. Orthop., *44*:205–220, 1966.

Chapter 12
NEW FLUORIDE AGENTS AND MODALITIES OF DELIVERY

John W. Stamm

Dean's discovery that fluoridated drinking water brought about a 50% reduction in dental caries experience among children led to two significant developments in dentistry. The first development was the 1945 introduction and subsequent expansion of community water fluoridation. As a result of this development, approximately 116 million people were served by fluoridated water in the United States by 1980,[1] and this number is still growing. The second initiative comprised research and development on topical fluorides that could be applied directly to erupted teeth for the prevention of caries. Since the early 1940s when this work began in earnest, considerable progress has been made in the formulation and delivery of efficacious topical fluoride preventives for clinical use in dentistry. In spite of this success, dental science has *not* become complacent. The search for better topical fluoride agents and more effective delivery vehicles continues relentlessly. It is the general goal of this chapter to provide an overview of some of the newer and/or more innovative topical fluorides that are emanating from dental research centers around the world.

Broadly speaking, innovations in topical fluoride therapy can be grouped into two categories. The first category comprises new fluoride compounds. In this chapter three of the newer fluoride chemicals will be discussed. The second category encompasses newer vehicles that actually deliver the fluoride agent to the enamel or root surface. The newer vehicles can in turn be subdivided into three traditional groupings: (1) self-applied fluorides; (2) professionally applied fluorides; and (3) certain dental materials. This chapter will review some recent developments in each of these three types of delivery mechanisms.

ALTERNATIVES TO CONVENTIONAL FLUORIDE COMPOUNDS

In North America, the conventional fluoride agents for topical fluoride therapy are stannous fluoride (SnF_2), sodium fluoride (NaF), acidulated phosphate fluoride (APF), and sodium monofluorophosphate (Na_2PO_3F). The use and efficacy of these caries-inhibiting compounds have long been established,[2-5] even though the precise biological and chemical mechanisms for caries inhibition are not completely understood.[6-7] Fluoride agents that are not commonly used in the United States are ammonium fluoride, amine fluoride, and titanium fluoride.

176

Ammonium Fluoride

Since elevated fluoride concentration in the tooth surface has been shown to be associated with lower caries experience,[8-9] the rationale for using ammonium fluorides in caries prevention is based upon the finding that ammonium fluoride (NH_4F) leads to greater fluoride uptake by enamel than conventional fluoride agents.[10-13] A partial explanation for the increased fluoride deposition may be related to the greater ability of ammonium fluoride to temporarily demineralize enamel, and thus promote fluoride deposition compared to sodium fluoride at the same pH. On a clinical level, however, a study by DePaola has been unable to demonstrate enhanced caries inhibition with ammonium fluoride rinses or dentifrices relative to conventional fluorides. Table 12–1 shows that *daily* use of high concentration sodium and ammonium fluoride rinses resulted in significant reductions in caries increment, but there was no clinical difference between the two fluoride rinses themselves. (Incidentally, the caries increment in this study seems inordinately high and it would be difficult to generalize from this investigation to other settings.) Based on all the clinical evidence available to date, there seems to be no advantage to recommending ammonium fluoride compounds for caries prevention. Indeed, taste considerations may make it a poorer choice than conventional fluoride agents.

Amine Fluoride

Amine fluorides are a group of organic fluoride formulations that have been extensively tested for caries prevention in Europe, particularly in Switzerland by Mühlemann and collaborators. Amine fluorides have two in vitro properties that make them attractive as potential caries preventives. These properties are: (1) enhanced fluoride uptake by enamel,[14-15] and (2) antibacterial activity against certain plaque microorganisms.[14,16-19] The precise mechanism of the antimicrobial effect of amine fluorides is not known. It does appear, however, to be more related to the organic portion of the molecule than due to the fluoride per se. This was confirmed by results from an animal experiment in which daily 15 min treatments with 250 ppm solutions of sodium fluoride, amine fluoride, amine chloride, and EHDP were applied to rodents on a high plaque-forming diet.

Turning to clinical considerations, amine fluorides have found considerable acceptance in Europe, largely through the efforts of the GABA company, which uses amine fluorides as the active ingredient in its Elmex dentifrices, topical fluoride gels, and mouthrinses. The available evidence from efficacy studies indicates, however, that while amine fluorides are probably as good as conventional fluorides, they are not clinically superior. For example, Table 12–2 shows that the unsupervised

Table 12–1. Effect of High Concentration Ammonium and Sodium Fluoride Rinses in 10- to 12-Year-Old Children after 24 Months of Use

Rinse	N	DFS Increment	% Reduction
NH_4F	159	4.40	41.7
NaF	158	4.43	41.3
Placebo	158	7.55	—

Source: De Paola, P.F., et al.: Effect of high-concentration ammonium and sodium fluoride rinses on dental caries in schoolchildren. Community Dent. Oral Epidemiol., 5:7–14, 1977. © 1977 Munksgaard International Publishers Ltd., Copenhagen, Denmark.

Table 12–2. Caries Inhibition by an Amine Fluoride Dentifrice in Grade 1–2 Children after 6 Years

Caries Index	Control Group, N = 59		Fluoride Group, N = 50		t-Test
	Mean	**S.D.**	**Mean**	**S.D.**	
$D_{3-4}FS$	8.39	5.77	5.62	5.46	P ≤ .05
$D_{1-4}FS$	18.34	10.53	13.90	10.04	P ≤ .05

Source: Marthaler, T.M.: Caries-inhibition by an amine fluoride dentifrice. Results after 6 years in children with low caries activity. Helv. Odontol. Acta, *18*(Suppl. 8):35–44, 1974.

home use of amine dentifrice over a 6-year period was efficacious in reducing caries increment,[20] but the caries savings obtained were almost identical to what would be expected from conventional fluoride dentifrices. In the United States, the amine fluoride dentifrice and the amine fluoride rinse have been shown to be as effective as, but not superior to, a stannous fluoride dentifrice and a sodium fluoride mouthrinse respectively. At the present time, therefore, it is reasonable to conclude that amine fluorides are as efficacious as conventional fluorides but that evidence for superior caries prevention is lacking.

Titanium Fluoride

In recent years reports have appeared that suggest that titanium fluoride may have considerable potential as a caries-preventing agent. In 1972 Shrestha et al.[21] and Mundorff et al.[22] showed that a 1% titanium tetrafluoride (TiF_4) solution was able to reduce enamel solubility. In 1976 Wei et al.[23] demonstrated that titanium tetrafluoride provided acceptable fluoride uptake in enamel and also formed a somewhat acid-resistant coating on the enamel surface. More recently, this work has been furthered by Wefel[24] and Clarkson and Wefel[25] who confirmed that application of titanium tetrafluoride in vitro resulted in less enamel fluoride uptake than occurs with APF solutions as shown in Figure 12–1. They also observed a tenacious, acid-resistant coating on the enamel. When subjected to an artificial caries lesion process, titanium tetrafluoride treated enamel proved to be more resistant to caries formation than the APF treated enamel. These results have stimulated investigations on titanium tetrafluoride uptake in cementum[26–27] and preliminary results are very consistent with in vitro enamel work.

To date there has only been limited clinical work with titanium tetrafluoride. Some of this research is shown in Table 12–3. Reed and Bibby[28] have reported that the annual, 1-min application of a 1% solution of titanium tetrafluoride to the teeth on one side of the mouth of each of 110 children, offered greater protection after 3 years than was provided for teeth on the other side of the mouth treated with a 4-min application of APF. Both treatments were effective when compared against an untreated control group comprising 88 children. One clinical study does not provide a sufficient basis on which to develop recommendations for clinical practice. It is hoped that more clinical work with titanium tetrafluoride will be carried out in the future.

ALTERNATIVES TO CONVENTIONAL FLUORIDE VEHICLES

Although there are many different ways in which fluoride agents may be brought to the tooth surface, it is useful to think of three general approaches. First, there are

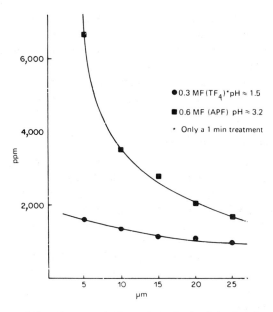

Fig. 12–1. Mean enamel fluoride concentrations at standardized depths after topical treatment with TiF$_4$ and APF. (From Wefel, J.S.: Artificial lesion formation and fluoride uptake after TiF$_4$ application. Caries Res., *16*:26–33, 1982.)

a whole host of procedures that rely on self-application. Second, there is the standard, classical method of professional application. Third, certain dental materials have long been known to transfer fluoride to the immediately adjacent tooth structure. There are new ideas and developments for fluoride delivery within all three of these broad categories.

The most conventional self-application procedure is the use of fluoride dentifrices. Rinsing with 0.2% NaF has become the next most common vehicle for self-applied fluorides. Two rather *unconventional* fluoride vehicles, fluoride-impregnated dental floss and fluoride-containing chewing gums, have received attention in recent years.

Fluoride-Impregnated Dental Floss

That floss impregnated with a fluoride compound could be used to increase the deposition of fluoride on interproximal tooth surfaces in vitro was demonstrated by Gillings in 1973.[29] This finding was confirmed by Bohrer et al.[30] as well as Chaet and Wei.[31] The latter investigators also showed that use of a fluoride-impregnated floss in vivo could reduce the number of interproximal sites demonstrating *Strep-*

Table 12–3. Effect of TiF$_4$ on 3-Year Caries Increment among 11 to 13-Year-Old Children

Agent	N	Baseline DMFS	DMFS Increment
TiF$_4$	110	3.85	3.10
APF	110	3.70	4.60
None	88	3.50	5.80

Source: Reed, A.J., and Bibby, B.G.: Preliminary report on effect of topical applications of titanium tetrafluoride on dental caries. J. Dent. Res., 55:357–358, 1976.

tococcus mutans. Further work on developing floss to deliver fluoride interproximally has been reported by Kaufman et al.,[32] Tsao et al.,[33] and Gellens et al.[34] These investigators have employed stannous fluoride in conjunction with floss and have reported reduced bacterial activity in supragingival plaque in vitro as well as increased fluoride concentrations in supragingival plaque in vivo. To date, no clinical studies have demonstrated that fluoride-containing floss prevents dental caries or inhibits gingivitis.

Fluoride-Containing Chewing Gum

There has been relatively little research on fluoride-containing chewing gums. One recent clinical study conducted in Thailand under the general supervision of the World Health Organization, however, compared the preventive efficacy of two fluoride-containing chewing gums with biweekly 0.2% NaF mouthrinsing.[35] Besides containing 0.55 mg F per stick of gum, the sweetener in one chewing gum comprised 30% saccharose, 30% dextrose, as well as 20% glucose; the sweetener in the second gum consisted of equal parts of xylitol and sorbitol. During school periods the children chewed four sticks of chewing gum per day over a 3-year period. The results from this study appear in Table 12–4. While it appears that the fluoride/xylitol combination was the most efficacious, followed by the fluoride mouthrinse, it is also apparent that this study experienced considerable difficulties during all phases of its implementation.

Professionally Applied Topical Fluorides

Most conventional topical fluorides intended for professional application consist of either solutions or gels. One undesirable characteristic of fluoride solutions and gels is that they remain present on the tooth surface for exceedingly short periods of time, thereby limiting the opportunity for the fluoride to react with enamel crystals. One significant strategy has been to develop controlled or slow release technologies, which can prolong the fluoride-enamel interaction from minutes to hours, weeks, or even months. Three significant advances in this direction are: (1) fluoride varnishes; (2) fluoride-containing resins for topical application; and (3) intraoral controlled release devices intended to supply 1 mg F/day for periods up to 6 months.

Fluoride Varnishes

One way of improving a topical fluoride agent's ability to increase the fluoride uptake in the enamel surface is to prevent the washing off of the agent, thereby increasing the fluoride's reaction time with enamel crystals. Richardson,[36] in an in

Table 12–4. Effect of F-Containing Chewing Gum on 3-Year Caries Increment among 7-Year-Old Children

Agent	N	DMFS		
		Baseline	Final	Increment
F rinse	205	1.66	2.73	1.07
F gum	174	1.42	3.91	1.49
F gum + xylitol	316	0.78	1.72	0.94

Source: Khambanonda, S., et al.: Prévention de la carie dentaire en Thailande: Trois produits flourés soumis à des tests comparatifs. J. Biol. Buccale, *11*:255–263, 1983.

Table 12–5. Fluoride Concentrations in Homologous Premolars 5 Weeks after Fluoride Varnish Treatment[39]

Group	N	Mean Biopsy Depth (μm)	Mean Fluoride Concentration (ppm)	Std. Error
Treatment	35	10.9	1,203	103.8
Control	35	12.3	612	57.1

Paired t-test: t = 8.27,

vitro study, has shown that lengthening the time interval between the topical fluoride treatment and the commencement of the washing action significantly increases fluoride uptake. This is why dental professionals ask patients to refrain from eating or ingesting liquids for at least 30 min following topical fluoride therapy. Of course, saliva cannot be controlled in this way. To overcome this, Richardson applied a waterproof coating on the teeth immediately after the fluoride application, temporarily sealing the fluoride to the tooth. His findings indicated that in vitro this approach was effective in increasing fluoride uptake by enamel. Evidently, the major drawback to this procedure was the impracticality of the method in the clinical setting. One approach to overcoming this obstacle was to incorporate the fluoride compound directly into a varnish-like coating material which, once applied, will adhere to the teeth for up to several hours. At least two such fluoride varnishes have been developed.

The first agent, Duraphat, is a rather viscous material that contains 50 mg NaF/ml or a concentration of 2.2% F ion. Developed in Germany by Schmidt,[37] Duraphat is applied to clean, dry teeth with an applicator, whereupon the varnish sets rapidly even in the presence of moisture. The varnish film, which is semipermeable to moisture, remains on the teeth for up to 12 hours during which time fluoride is continuously released to react with enamel. The second agent, called FluorProtector, is a thinner, polyurethane lacquer containing 1% difluorosilane.[38] This material is applied to clean, dry teeth with a small specially provided brush, whereupon the varnish dries rapidly to a thin, completely transparent coating. Dryness of the teeth must be maintained during varnish application.

In terms of fluoride uptake by enamel, both agents have demonstrated impressive in vivo results. Table 12–5 shows that 5 weeks after Duraphat application, the fluoride in the outer enamel is still double the concentration in the control teeth.[39] This result has been confirmed and extended to include difluorosilane by more recent studies.[40–41] Fluoride uptake, however, is not a proxy measure of caries reduction and hence one has to look to properly conducted clinical trials to judge the usefulness

Table 12–6. The Preventive Effects of Fluoride Varnishes

Study	Length of Study	Number of Applications	% Reduction DMFS	% Reduction DMFT
Maiwald & Geiger[42]	2 years	6	45	
Hetzer & Irmisch[43]	3 years	5	43	
Lieser & Schmidt[44]	3 years (urban)	5–6		62
	3 years (rural)	5–6		48

Table 12–7.   Mean DMFS Increments for Continuous Participants after 20 Months[45]

Groups	N	12 Months	20 Months	20 Months % Reduction
FluorProtector	201	0.89	1.70	15.8
Duraphat	255	0.98	1.73	14.4
Control	247	0.96	2.02	—

of topical fluoride varnishes in preventive dentistry. Such clinical studies are still relatively few in number. Table 12–6 shows some caries reductions in children whose initial age varied between 9 and 12 years.[42–44] Table 12–7 shows some interim results from the only clinical study with fluoride varnish underway in North America.[45] It is worth noting that the latter results appear to be considerably lower than those from the European studies.

A good review on topical fluoride varnishes and their effect on caries has been prepared by Clark.[45] Other recent studies appear in Seppä et al.[46] and Seppä.[47] Finally, fluoride varnishes are also potentially useful in the treatment of dentinal hypersensitivity.[48] Neither Duraphat nor FluorProtector have been approved by the FDA or the ADA Council on Dental Therapeutics, so their use in the United States must await application to and concurrence from these agencies.

Fluoride-Exchanging Resins

A completely different vehicle for topical fluoride application was proposed by Rawls and Querens in 1980.[49] These authors suggested that anion-exchange resins possess the potential for increasing fluoride uptake by enamel. More recently, Rawls and Zimmerman[50] have shown that fluoride-exchange resins are particularly effective in penetrating the demineralized enamel in early carious lesions produced with the artificial caries model. Once the resin deposits itself in the porous enamel, it sets and begins to release fluoride ions in controlled amounts to the local environment. Early in vitro experiments suggest that this new technology has potential. The research, however, is still in its preliminary stages and will require much more laboratory work followed by clinical research before fluoride-exchanging resins will find a place in dental practice.

Intraoral Controlled Release Devices for Fluorides

Since 1980 there has been considerable interest in the clinical application of intraoral controlled release devices for fluorides (CFRD). While essentially a method for delivering systemic fluorides, the fact is that intraoral CFRDs appreciably elevate the salivary fluoride levels and can thus be expected to contribute to a topical fluoride effect. While CFRDs represent the result of complex chemistry, the principle of these devices is relatively simple. Essentially, a fluoride-containing copolymer matrix serves as a reservoir for fluoride. This reservoir is sandwiched or encapsulated by a copolymer membrane that controls the rate at which fluoride ions from the reservoir are able to migrate out into the oral environment.

The controlled fluoride-release devices can be extremely accurate over prolonged periods in their ability to release a predetermined quantity of fluoride. For example, they have been shown to release precisely 1 mg F/day over a period of 140 days.

An example of a practical benefit for the CFRD is that it can overcome the serum fluoride spiking brought about by the ingestion of fluoride supplements. By using a CFRD, a lower quantity of fluoride is released throughout the day resulting in a much more even and physiologic fluoride concentration in the serum.

The effect of controlled fluoride-release devices on salivary fluoride concentration can be observed in Figure 12–2, which was taken from Mirth.[51] Prior to the intraoral attachment of the CFRD the mean salivary fluoride concentration is around 0.02 ppm F. As soon as the device is placed in the mouth, however, the salivary fluoride concentration rises and remains relatively constant for up to 100 days. This elevated salivary fluoride concentration has a predictable effect on the in vivo plaque fluoride levels. Figure 12–3 taken from Mirth et al.[52] shows that in the 29 days before the CFRD is installed, the plaque fluoride levels remain between 10 and 15 ppm in the upper and lower teeth. Once the CFRD is attached in the mouth the plaque fluoride concentration jumps dramatically, reflecting the elevated salivary fluoride levels. Once the device is removed, plaque fluoride concentrations tend to fall back to their baseline.

What is the practical or clinical significance of the controlled fluoride-release devices? In my opinion their use is extremely limited in spite of their technical elegance. The CFRDs measure approximately 8×3×2 mm and two such devices must be attached simultaneously, usually on the buccal surface of the upper first permanent molars. In the one clinical evaluation of the CFRDs, involving 11 adults for 35 days, many successful outcomes were registered. There were, however, a sufficient number of genuine problems to cast a shadow over the general utility of this method for

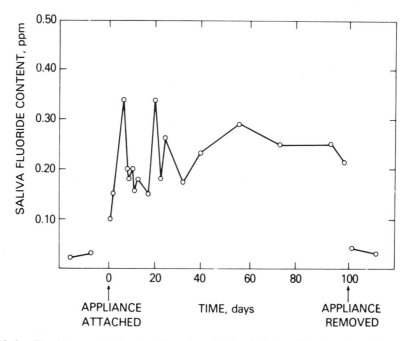

Fig. 12–2. Fluoride concentration in pilocarpine-stimulated whole saliva from beagle dog wearing intraoral controlled fluoride release system providing 0.5 mg F/day. (From Mirth, D.B.: The use of controlled and sustained release agents in dentistry: A review of applications for the control of dental caries. Pharmacol. Ther. Dent., 5:59–67, 1980.)

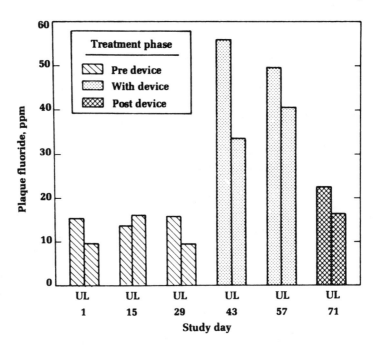

Fig. 12–3. Mean maxillary (U) and mandibular (L) fluoride concentrations (wet weight) in persons wearing intraoral controlled fluoride release devices.[52]

providing a combination of systemic and topical fluorides. Table 12–8 shows what transpired in just over 1 month with the adults. What the situation would have been had these devices been attached in the mouths of children is not known. Yet it is among children where caries prevention is most critical. It seems reasonable to suggest that unless dramatic improvements were to occur in the CFRDs, their application to caries prevention remains in serious doubt.

Table 12–8. Controlled Fluoride Release Device

11 Patients—2 Devices Each
2 Patients Lost a Device within 35 Days
3 Patients Had a Device Removed within 2, 2, and 20 Days
5 of 22 Devices Caused Mucosal Irritation

FLUORIDE IN DENTAL MATERIALS

Dental materials represent a third category of fluoride vehicles, insofar as certain types of dental materials either contain high concentrations of fluoride or lend themselves to the addition of fluoride. For example, although they have only limited use today, silicate cements were at one time the material of choice for anterior esthetic restorations. Although silicate cements were known to have some suboptimal physical properties, the fact that the silicate restorations transferred a considerable amount of fluoride to the tooth structure immediately adjacent to the filling resulted in a low occurrence of secondary caries.[53] Beyond silicates, the current interest in dental materials as fluoride vehicles focuses on three types of products, which are: (1) alginates; (2) cements and luting agents; and (3) amalgams.

Alginates

In 1978, Hattab and Frostell reported that alginate impression materials contain high levels of fluoride and are capable of releasing this fluoride in ionic form to the immediately surrounding environment.[54] Various fluoride salts are added to alginates to ensure a casting with a hard dense surface and to act as an accelerator during the setting of certain gypsum products.[55] Table 12–9 shows the fluoride concentration in two alginate impression materials available in Sweden. Table 12–10 demonstrates that when an impression is taken with an alginate material (Zelgan), there is appreciable fluoride uptake by tooth enamel. It has also been shown that a significant increase in a patient's serum fluoride level occurs within an hour of taking the impression.[56–57] Although it is felt by some that the use of alginates as vehicles for professionally applied fluorides offers better intraoral control of the fluoride agent,[58] it has yet to be shown that this is an economical and clinically efficacious method for topical fluoride application, or indeed, caries prevention.

Table 12–9. Fluoride Concentration in Different Weights of Alginate Powder*

	Weight of the Sample (mg)	Fluoride Concentration (ppm)
Zelgan	90.5	19,890
	67.0	17,612
	53.4	17,978
	28.0	19,286
$\bar{X} \pm$ S.D.		18,692 ± 1,075
Kerr	97.3	13,690
	60.5	14,546
	33.8	15,385
	32.9	14,894
$\bar{X} \pm$ S.D.		14,629 ± 714

Table 12–10. Fluoride Concentration (ppm) in Buccal Surface of Homologous Molars after 5 Minute Exposure to Alginate Gel*

Layer	Test Mean	S.E.	Control Mean	S.E.	Fluoride Uptake	P Value
1	1,421	167	807	57	614	≤.05
2	748	64	608	47	140	≥.1

*Source: Hattab, F., and Frostell, G.: The release of fluoride from two products of alginate impression materials. Acta Odontol. Scand., *38*:385–395, 1980.

Cements

Although the role of fluoride in esthetic silicate restorations has been known for some time, serious interest in other dental materials as fluoride vehicles is relatively recent. As a first step, LeGrand et al. analyzed the fluoride concentrations in a number of normal dental restorative materials available in Switzerland.[59] Their results are presented in Figure 12–4. Not surprisingly, the silicate materials contain the highest concentration of fluoride. Also interesting is that one composite restorative

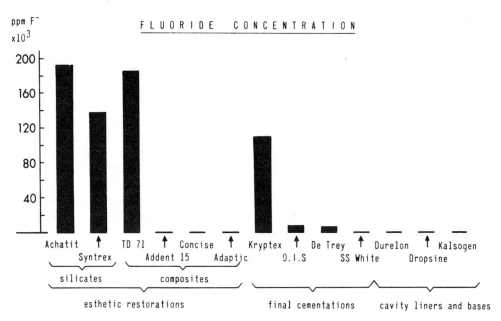

Fig. 12–4.   Fluoride concentration in various dental materials. (From Legrand, M., de Carlini, C.H., and Cimasoni, G.: Fluoride content and liberation from dental cements and filling materials. Helv. Odontol. Acta, *18*:114–118, 1974.)

material (TD71, Dental Fillings Ltd., London) and one dental cement (Kryptex, SS-White Co., Philadelphia) contain elevated levels of fluoride. Since the presence of fluorides in luting cements does not affect their essential physical properties, fluorides have been successfully added to polycarboxylate (Poly-F, DeTrey Frères, Zurich) and glass ionomer (ASPA, Amalgamated Dental, London) cements.[60–61] Table 12–11 indicates that fluoride-containing polycarboxylate cement can release fluoride to hydroxyapatite powder as rapidly and readily as can a silico-phosphate cement.[60] Fluoride-containing glass ionomer cement, being somewhat less soluble, releases its fluoride more slowly.[61] Clinically, because luting cements are present in only a thin film, the actual quantity of fluoride released is small and poses no threat to the health of pulpal tissues.

Amalgams

Since amalgam restorations form such an important part of clinical dental practice[62] and because of the very real problems with recurrent decay around margins,[63] interest

Table 12–11.   Fluoride Increase (μg/mg) in Hydroxyapatite from Cement Specimen Immersed in 0.01 M Phosphate Buffer Solution

Week	Silico-phosphate		Poly-carboxylate	
	Mean	S.D.	Mean	S.D.
1	.58	.18	.64	.13
2	.11	.02	.11	.01
3 + 4	.20	.05	.25	.06
5	.07	.02	.06	.03

Source: Forsten, L.: Fluoride release from a fluoride-containing amalgam and two luting cements. Scand. J. Dent. Res., *84*:348–350, 1976.
© 1976 Munksgaard International Publishers Ltd., Copenhagen, Denmark.

Table 12–12. Fluoride Concentration in Enamel and Dentin Adjacent to F-Containing and Conventional Amalgams after 7 Weeks in an Artificial Saliva Environment

		Mean F Concentration (ppm)	
Tooth Structure	Depth μm	Fluoride Amalgam n = 12	Non-F Amalgam n = 6
Enamel	.05	4,200	1,300
	3.3	2,600	700
	8.6	3,000	800
Dentin	.06	3,600	1,800
	4.5	6,500	2,600
	11.5	9,000	2,700

Source: Tveit, A.B., and Lindh, U.: Fluoride uptake in enamel and dentin surfaces exposed to a fluoride-containing amalgam in vitro. Acta Odontol. Scand., *38*:279–283, 1980.

has developed for the incorporation of fluorides into amalgam restorative materials. A number of recent studies have examined the in vitro properties of fluoride-containing amalgams. A particularly good overview has been provided by Hurst and von Fraunhofer.[64] These authors tested Amalcap, Aristaloy, and Dispersalloy in their nonfluoride and fluoridated formulations. Also evaluated was Yalta alloy, a Japanese product, which was only available in a fluoride-containing form.

Fluoride-containing amalgams must meet a large range of criteria to be acceptable for clinical dental practice. To be considered at all, the set amalgam must be capable of releasing its fluoride to the immediately adjacent tooth structure. In this respect, the in vitro work reported by Tveit and Lindh[65] appears to be encouraging, as shown in Table 12–12. At every depth of a preparation filled with 1% Fluoralloy, the fluoride concentration in the tooth structure was higher than in a preparation on the same tooth filled with a conventional alloy.

Even more important for clinical dental practice, however, are the alloy's corrosion resistance and compressive strength. Hurst and Fraunhofer have demonstrated that fluoride-containing amalgams have higher corrosion activity than do the nonfluoridated control restorations.[64] More seriously, fluoride in amalgam appears to reduce significantly the compressive strengths of amalgam. Table 12–13 shows clearly that in set, noncorroded amalgam, fluoride caused a significant drop in compressive strength. When amalgam is subjected to corrosion, compressive strength is slightly reduced in nonfluoridated amalgams while it drops significantly in the fluoride-containing restorations, particularly Amalcap and Aristaloy. Based on these results,

Table 12–13. Compressive Strength of Fluoride-Containing Amalgams

	Set Amalgam		Corroded Amalgam	
	Control	F	Control	F
Amalcap	522.0	451.5	498.8	362.0
Aristaloy	549.0	473.5	533.8	388.3
Dispersalloy	604.8	556.3	597.5	546.8
Yalta F	—	550.8	—	503.8

Source: Hurst, P.S., and von Fraunhofer, J.A.: In vitro studies of fluoridated amalgam. J. Oral Rehabil., *5*:51–62, 1978.

and notwithstanding an early clinical report by Minoguchi et al.,[66] fluoride-containing amalgams cannot be recommended for dental practice at this time. More research is required to improve their physical properties before such products can be considered for use in the dental office.

CONCLUDING REMARKS

Four general topics have been covered. First, I reviewed three nonconventional (in the U.S. context) fluoride compounds: ammonium fluoride, amine fluoride, and titanium tetrafluoride. It was suggested that the first two do not demonstrate clear superiority over conventional agents but that research on titanium tetrafluoride should be pursued especially if it can play a role in root caries prevention.

Second, I reviewed dental floss and chewing gum as vehicles for self-applied fluorides. More investigations will be necessary before these products can claim a role in preventive dentistry, but the fact that floss, like other interdental cleaners, acts interproximally is an attractive concept.

Third, I discussed three types of slow fluoride release products. These were fluoride varnishes, fluoride-containing resins, and controlled fluoride-release devices. At this time, fluoride varnishes are not available for clinical use in the United States. They are, however, fairly well tested and are as effective as other professionally applied fluoride agents. Nevertheless, the advantages to fluoride varnish may be that it allows a faster fluoride treatment. It may also be efficacious in preventing dentinal hypersensitivity. Fluoride-exchange resins are very new and have not been clinically tested. The CFRDs, in my view, have little general application in preventive dentistry.

Fourth, I reviewed alginates, cements, and amalgam materials as vehicles for fluoride delivery. Of these, only fluoride-containing luting cements seem to have a place in clinical dental practice at this time.

REFERENCES

1. Fluoridation Census 1980, US Department of Health and Human Services, US Public Health Service, Center for Disease Control, Atlanta.
2. Horowitz, H.S.: Review of topical applications. Fluorides and fissure sealants. J. Can. Dent. Assoc., 46:38–42, 1980.
3. Mellberg, J.R., and Ripa, L.W.: Fluoride in Preventive Dentistry: Theory and Clinical Applications. Chicago, Quintessence Publishing Co., 1983.
4. Murray, J.J., and Rygg-Gunn, A.J.: Fluorides in Caries Prevention. 2nd Ed. Bristol, Wright, 1982.
5. Newbrun, E.: Fluorides and Dental Caries. Springfield, Charles C Thomas, 1972.
6. Weatherell, J.A., Deutsch, D., Robinson, C., and Hallsworth, A.S.: Assimilation of fluoride by enamel throughout the life of the tooth. Caries Res., 11(Suppl. 1):85–101, 1977.
7. Driessens, F.C.M.: Mineral aspects of dentistry. Monogr. Oral Sci., 10:129–142, 1982.
8. Richards, A.: Fluoride content of buccal surface enamel and its relation to dental caries in children. Arch. Oral Biol., 22:425–428, 1977.
9. Bookstein, F.L., and DePaola, P.F.: A potential model for the interaction of enamel fluoride and plaque in the development of dental caries. J. Dent. Res., 56:40–45, 1977.
10. Caslavska, V., Brudevold, F., Vrbic, V., and Moreno, E.C.: Response of human enamel to topical application of ammonium fluoride. Arch. Oral Biol., 16:1173–1180, 1971.
11. Grøn, P., and Caslavska, V.: Fluoride deposition in enamel from application of sodium, potassium or ammonium fluoride. Caries Res., 15:459–467, 1981.
12. Caslavska, V., Grøn, P., Stern, D., and Skobe, Z.: Chemical and morphological aspects of fluoride acquisition by enamel from topical application of ammonium fluoride with ammonium monofluorophosphate. Caries Res., 16:170–178, 1982.
13. DePaola, P.F., et al.: Effect of high-concentration ammonium and sodium fluoride rinses on dental caries in schoolchildren. Community Dent. Oral Epidemiol., 5:7–14, 1977.

14. Klimek, J., Hellwig, E., and Ahrens, G.: Fluoride taken up by plaque, by the underlying enamel, and by clean enamel from three fluoride compounds in vitro. Caries Res., *16*:156–161, 1982.
15. Mühlemann, H.R., Schmid, H., and König, K.G.: Enamel solubility reduction studies with inorganic and organic fluorides. Helv. Odontol. Acta, *1*:23–33, 1957.
16. Bramstedt, F., and Bandilla, J.: Ubet den Einfluss organischer Fluorverbindungen auf Säurebildung und Polysaccharidsynthese von Plaque-Streptokokken. Dtsch. Zahnaerztl. Z., *21*:1390–1396, 1966.
17. Shern, R., Swing, K.W., and Crawford, J.J.: Prevention of plaque formation by organic fluorides. J. Oral Med., *25*:93–97, 1970.
18. Shern, R.J., Rundell, B.B., and Defever, C.J.: Effects of an amine fluoride mouthrinse on the formation and microbial content of plaque. Helv. Odontol. Acta, *18*(Suppl. 8):57–62, 1974.
19. Schneider, P.H., and Mühleman, H.R.: The antiglycolytic action of amine fluorides on dental plaque. Helv. Odontol. Acta, *18*(Suppl. 8):63–70, 1974.
20. Marthaler, T.M.: Caries-inhibition by an amine fluoride dentifrice. Results after 6 years in children with low caries activity. Helv. Odontol. Acta, *18*(Suppl. 8):35–44, 1974.
21. Shresta, B.M., Mundorff, S.A., and Bibby, B.G.: Enamel dissolution. I. Effects of various agents and titanium tetrafluoride. J. Dent. Res., *51*:1561–1566, 1972.
22. Mundorff, S.A., Little, M.F., and Bibby, B.G.: Enamel dissolution. II. Action of titanium tetrafluoride. J. Dent. Res., *51*:1567–1571, 1972.
23. Wei, S.H., Soboroff, D.M., and Wefel, J.S.: Effects of titanium tetrafluoride on human enamel. J. Dent. Res., *55*:426–431, 1976.
24. Wefel, J.S.: Artificial lesion formation and fluoride uptake after TiF₄ application. Caries Res., *16*:26–33, 1982.
25. Clarkson, B.H., and Wefel, J.S.: Titanium and fluoride concentrations in titanium tetrafluoride and APF treated enamel. J. Dent. Res., *58*:600–603, 1979.
26. Hals, E., Tveit, A.B., Totdal, B., and Isrenn, R.: Effect of NaF, TiF₄, and APF solutions on root surfaces in vitro, with special reference to uptake of F. Caries Res., *15*:468–476, 1981.
27. Tveit, A.B., et al.: Fluoride uptake by dentin surfaces following topical application of TiF₄, NaF, and fluoride varnishes in vivo. J. Dent. Res., *63*:(Abstract 1313), 1984.
28. Reed, A.J., and Bibby, B.G.: Preliminary report on effect of topical applications of titanium tetrafluoride on dental caries. J. Dent. Res., *55*:357–358, 1976.
29. Gillings, B.R.: Fluoride uptake of enamel after application of fluoride solutions and fluoride-impregnated dental floss: A preliminary report. J. Dent. Res., *52*(Abstract 33):575, 1973.
30. Bohrer, D., Hirschfeld, Z., and Gedalia, I.: Fluoride uptake in vitro by interproximal enamel from dental floss impregnated with amine fluoride gel. J. Dent., *11*:271–273, 1983.
31. Chaet, R., and Wei, S.H.: The effect of fluoride impregnated dental floss on enamel fluoride uptake in vitro and *Streptococcus mutans* colonization in vivo. J. Dent. Child., *44*:122–126, 1977.
32. Kaufman, A.K., et al.: Physical and antibacterial properties of SnF₂-coated dental floss. J. Dent. Res., *61*:(Abstract 1219), 1982.
33. Tsao, T.F., et al.: Clinical, substantivity and microbiological studies of fluoride-coated dental floss. J. Dent. Res., *61*:(Abstract 1220), 1982.
34. Gellens, A.J., Kaufman, A.K., Newman, M.G., and Carranza, F.A.: Physical and antimicrobial studies of stannous fluoride-coated dental floss. J. Dent. Res., *62*:(Abstract 72), 1983.
35. Khambanonda, S., et al.: Prévention de la carie dentaire en Thaïlande: Trois produits fluorés soumis à des tests comparatifs. J. Biol. Buccale, *11*:255–263, 1983.
36. Richardson, B.: Fixation of topically applied fluoride in enamel. J. Dent. Res., *46*:(Suppl.):87–91, 1967.
37. Schmidt, H.F.: Ein neues Touchierungsmittel mit besonders lang anhaltendem intensivem Fluoridierungseffekt. Stomatol. DDR., *17*:14, 1964.
38. Arends, J., and Koulourides, T.: The effect of silane fluoride, NaF and SnF₂ on cariogenicity. J. Dent. Res., *56*:(Abstract 268), 1977.
39. Stamm, J.W.: Fluoride uptake from topical sodium fluoride varnish measured by an in vivo enamel biopsy. J. Can. Dent. Assoc., *40*:501–505, 1974.
40. Petersson, L.G.: In vivo fluoride uptake in human enamel following treatment with a varnish containing sodium fluoride. Odont. Revy, *26*:253–266, 1975.
41. Seppä, L., Luoma, H., and Hausen, H.: Fluoride content in enamel after repeated applications of fluoride varnishes in a community with fluoridated water. Caries Res., *16*:7–11, 1982.
42. Maiwald, H.J., and Geiger, L.: Lokalapplikation von Fluorschutzlack zur Kariespraevention in Kallektiven nach dreijahriger Kontrollzeit. Stomatol. DDR, *24*:123–125, 1973.
43. Hetzer, B., and Irmisch, B.: Kariesprotektion durch Fluorlack (Duraphat), Klinische Ergebnisse und Erfahrungen. Dtsch. Stomatol. Z., *23*:917–922, 1973.
44. Lieser, O., and Schmidt, H.F.: Caries preventive effect of fluoride lacquer after several years' use in children. Dtsch. Zahnaerztl. Z., *33*:176–178, 1978.
45. Clark, D.C.: A review on fluoride varnishes: An alternative topical fluoride treatment. Community Dent. Oral Epidemiol., *10*:117–123, 1982.

46. Seppä, L., Hausen, H., and Luoma, H.: Relationship between caries and fluoride uptake by enamel from two fluoride varnishes in a community with fluoridated water. Caries Res., *16*:404–412, 1982.

47. Seppä, L.: Effect of dental plaque on fluoride uptake by enamel from a sodium fluoride varnish in vivo. Caries Res., *17*:71–75, 1983.

48. Clark, D.C., Hanley, J.A., Weinstein, P.L., and Stamm, J.W.: An empirically based system to estimate the effectiveness of caries preventive agents. Caries Res., (in press).

49. Rawls, H.R., and Querens, A.E.: The potential of an adhesive anion-exchange resin as a fluoride-releasing sealant. J. Dent. Res., *59*:(Abstract 895), 1980.

50. Rawls, H.R., and Zimmerman, B.F.: Fluoride-exchanging resins for caries protection. Caries Res., *17*:32–43, 1983.

51. Mirth, D.B.: The use of controlled and sustained release agents in dentistry: A review of applications for the control of dental caries. Pharmacol. Ther. Dent., *5*:59–67, 1980.

52. Mirth, D.B., et al.: Clinical evaluation of an intraoral device for the controlled release of fluoride. J. Am. Dent. Assoc., *105*:791–797, 1982.

53. Wilson, A.D., and Batchelar, R.F.: Dental silicate cements. I. The chemistry of erosion. J. Dent. Res., *46*:1076–1085, 1967.

54. Hattab, F., and Frostell, G.: The release of fluoride from alginate impression materials. Community Dent. Oral Epidemiol., *6*:273–274, 1978.

55. Hattab, F., and Frostell, G.: The release of fluoride from two products of alginate impression materials. Acta Odontol. Scand., *38*:385–395, 1980.

56. Whitford, G.M., and Ekstrand, J.: Systemic absorption of fluoride from alginate impression material in humans. J. Dent. Res., *59*:782–785, 1980.

57. Hattab, F.: Absorption of fluoride following inhalation and ingestion of alginate impression materials. Pharmacol. Ther. Dent., *6*:79–86, 1981.

58. Hattab, F.: Studies on alginates as vehicles for topical fluoride application. Karolinska Institute, Stockholm, 1983.

59. Legrand, M., de Carlini, C.H., and Cimasoni, G.: Fluoride content and liberation from dental cements and filling materials. Helv. Odontol. Acta, *18*:114–118, 1974.

60. Forsten, L.: Fluoride release from a fluoride-containing amalgam and two luting cements. Scand. J. Dent. Res., *84*:348–350, 1976.

61. Tveit, A.B., and Gjerdet, N.R.: Fluoride release from a fluoride-containing amalgam, a glass ionomer cement, and a silicate cement in artificial saliva. J. Oral Rehabil., *8*:237–241, 1981.

62. Skinner, E.W.: Science of Dental Materials. Philadelphia, W.B. Saunders, 1973.

63. Richardson, A.S., and Boyd, M.A.: Replacement of silver amalgam restorations by 50 dentists during 246 working days. J. Can. Dent. Assoc., *39*:556–559, 1973.

64. Hurst, P.S., and von Fraunhofer, J.A.: In vitro studies of fluoridated amalgam. J. Oral Rehabil., *5*:51–62, 1978.

65. Tveit, A.B., and Lindh, U.: Fluoride uptake in enamel and dentin surfaces exposed to a fluoride-containing amalgam in vitro. Acta Odontol. Scand., *38*:279–283, 1980.

66. Minoguchi, G., Tani, Y., and Tamai, S.: Abhandlung über die zinnfluoride enthaltenden Silberamalgame and über die Rolle, die sie als vorbeugende Mittel gegen die Zahnkaries spielen. Fachblatt für Stomat., *7*:1–23, 1967.

Section IV

CARL A. VERRUSIO, MODERATOR

Chapter 13
CLINICAL USES OF FLUORIDE FOR THE SPECIAL PATIENT

James J. Crall and Arthur J. Nowak

This chapter focuses on the clinical uses of fluoride for those patients who, because of an existing physical or mental condition, are at an increased risk to develop dental disease or are unable to readily take advantage of conventional forms of fluoride therapy. The discussion includes identification of special populations such as: the developmentally disabled, chronically ill, immunodeficient or immunosuppressed, elderly, and immigrant populations. This chapter also presents some recommended alternatives to conventional therapy for these special populations.

Predisposing factors to be considered should include atypical morphology, physical limitations, mental limitations, social and societal influences, dietary modifications, and medical therapy. Other special considerations are accessibility, ability to cooperate, tissue irritation, and toxicity of the agents.

Numerous studies have provided evidence of a decline in caries prevalence in children and young adults in developed countries including the United States in recent years.[1-4] Increased availability and use of systemic and topical fluorides are regarded as the most important factors responsible for this decline.[5] As König pointed out at a recent conference in Zurich, however, efforts to implement prevention have been mainly directed to the community and group level (i.e., water fluoridation and school programs).[6] He further emphasized that in spite of the progress made at this level, many groups and individuals are at risk for which the "right approach" has not been defined or elaborated upon. This chapter will focus on the clinical uses of fluoride for those individuals who, because of an existing physical, mental or environmental condition, are at an increased risk to develop dental disease or are unable to readily take advantage of traditional forms of fluoride therapy. In particular, the discussion will center around the following special populations: (1) the developmentally disabled; (2) the chronically ill; (3) individuals who are immunodeficient or immunosuppressed; (4) the elderly; and (5) immigrant populations.

PREVALENCE OF DENTAL DISEASES IN SPECIAL POPULATIONS

The term developmentally disabled primarily denotes individuals with a disability attributable to mental retardation, cerebral palsy, epilepsy, autism or some form of learning impairment. The prevalence of these conditions has generally been estimated to be approximately 10% of a population.[7]

193

Several reports exist on the prevalence of dental diseases in the handicapped population. Although there is generally agreement as to a high prevalence of periodontal disease, malocclusion, and poor oral hygiene, the evidence regarding the prevalence and incidence of dental caries is conflicting. Much of the disagreement stems from the residence of the population being examined. Whether a person is institutionalized or noninstitutionalized appears to make a considerable difference with regard to their daily care including personal hygiene and diet. Unfortunately, most reports in the literature are based on studies that were conducted in large institutions, generally located in nonfluoridated areas. In most cases, radiographs were not available for interproximal caries detection. Finally, most of the studies were completed in the 1960s or early 1970s, prior to the advent of deinstitutionalization and normalization programs.[8]

Nowak recently reported on the caries prevalence of a large noninstitutionalized handicapped population.[8] These individuals were part of an outreach program established by the National Foundation of Dentistry for the Handicapped to provide daily preventive care, dental screenings, and appropriate referral for dental treatment for this segment of the population. Data obtained from 3,622 handicapped subjects showed that mean DMFT values for handicapped individuals were similar to values for nonhandicapped controls of similar ages obtained from a recent national survey.[9] Caries rates were generally slightly lower in subjects residing in fluoridated areas. Comparisons of DMFT values for specific handicapping conditions including mental retardation, cerebral palsy, and Down's syndrome indicated similar prevalence profiles for these groups.

Individuals with chronic metabolic and systemic disorders constitute another significant segment of the population whose numbers are increasing as medical technology advances. Chronic conditions that have been reported to be associated with an increased caries rate include congenital heart disease, uncontrolled diabetes mellitus, hypopituitarism, hypoparathyroidism, and drug abuse.[10] Special attention to prevention is also warranted for those with medical conditions wherein possible sequela to dental caries could lead to significant medical problems (i.e., those with cardiac problems, hemophilia, leukemia or sickle cell anemia). Soft tissue lesions, periodontal disease, and malocclusion are also reported to be among the oral manifestations of many systemic and metabolic disorders.

A third general group reported to be at an increased risk for the development of caries includes those who are immunodeficient or immunosuppressed. Over 17 types of primary immunodeficiency have been identified, most of which are hereditary diseases. Legler et al., compared the caries experience of 45 immunodeficient subjects of various subgroup classifications with a group of matched controls.[11] They found that the immunodeficient groups had significantly higher DMFS scores, ranging from approximately 25 to 450% greater than the matched control groups. These differences were not related to previous fluoride experience or plaque scores. Immunodeficient individuals also exhibited a greater extent of periodontal disease.

Alongside those with primary immunodeficiency are an ever increasing number of individuals whose immune responses have been suppressed as a result of medical therapy. A partial listing of conditions that are treated with immunosuppressives includes arthritis, glomerulonephritis, lupus erythematosis, organ transplantation, and ulcerative colitis. Caries rates in these individuals have not been widely studied. Immunosuppression is also associated with cancer therapy, although caries in these individuals is usually thought to be related to changes brought about as a result of chemotherapy or radiation therapy.[12]

Another segment of the population whose numbers have steadily increased and who are projected to compose a growing proportion of the population are the elderly.[13] Epidemiological data indicate an increase in the mean number of teeth present in the adult population and that further reductions in the number of individuals who are edentulous can be anticipated as a result of advancements in dental care.[14]

Evidence seems to suggest that root caries is emerging as a considerable problem in adults, one which will undoubtedly warrant additional attention in the future. Table 13–1 summarizes some preliminary results obtained from a recent survey of 520 noninstitutionalized elderly persons living in Iowa. This table demonstrates that the mean number of teeth present in dentate individuals over the age of 65 was 18.7. Nearly 63% of those teeth had evidence of gingival recession and, therefore, were at risk to develop root caries. Nineteen percent of the subjects had new unfilled carious lesions on the crowns of their teeth, while 25% had new unfilled root surface lesions. Recurrent caries was noted on the crowns of 17% of the subjects and on the roots of 3.5% of the sample. Overall, 32% had unfilled coronal caries and 28% had unfilled root caries. These data demonstrate that the prevalence of root caries in this noninstitutionalized population was equal to the prevalence of coronal caries. Unfortunately, relatively little is known about the etiology, pathogenesis or prevention of root surface caries.

One final category of individuals who have been shown to exhibit significantly poorer dental health than the majority of their counterparts in the United States are immigrants. Pollick and coworkers recently reported on the results of screening examinations performed as part of the Bay Area Human Nutrition Center Newcomer Project.[15] Their results indicate overall, that 30% of all children examined had serious dental conditions at an initial examination and that the percentage remained unchanged at a follow-up conducted 6 months later. Another 30% had nonserious conditions that needed treatment. Considerable variation existed among the immigrants as indicated by Table 13–2. Also of interest was the observation that after 6 months, some of the groups showed improvements in dental health, while others exhibited deterioration. This increase in treatment needs may reflect an acceleration

Table 13–1.   65+ Oral Health Study Baseline Data, 1984

N = 520 Dentate persons age 65–96 representative of the noninstitutionalized population of 2 Iowa counties

(1) $\bar{x}$ number of teeth/person = 18.7
(2)   % of teeth with recession = 62.9

		Coronal	Root
(3)	% of people with		
	initial unfilled decay	18.8	24.6
	recurrent unfilled decay	17.3	3.5
	unfilled decay	32.1	27.7
	filled teeth	86.7	50.0
	decayed fillings	89.2	63.0
(4)	In persons with unfilled decay, $\bar{x}$ D teeth =	2.1	2.1
(5)	In dentate population, $\bar{x}$ D teeth =	0.7	0.7

Source: Beck, J.S., et al., unpublished data, 1984.

Table 13–2. Percentage of Immigrant Populations with Serious Dental Conditions*

Newcomer Group	Initial (%)	6 Months (%)
Tagalog	75	45
Chinese-speaking	30	40
Vietnamese-speaking	30	25
Spanish-speaking	20	25
Cambodian-speaking	20	8

*Source: Pollick, H.F., Rice, A.J., and Echenberg, D.: Utilization of dental services by recent immigrant school children. J. Dent. Res., *63*(Abstract 1127):296, 1984.

of the disease process as a result of increased refined carbohydrate intake. This phenomenon is not limited to immigrants in the United States, similar findings have also been reported for Finnish immigrants in Sweden.[16]

Predisposing Factors

No doubt, everyone is familiar with the general factors that have been implicated in the caries process. I would like to point out, however, an additional "layer" of complicating factors that may predispose individuals in the previously outlined special populations to dental caries. These include atypical composition, morphology and position of teeth, physical limitations that may interfere with preventive measures and promote food and plaque retention, mental limitations that may necessitate assistance from a second party, social and societal factors that may influence the individual's appreciation and demand for dental care; and finally, dietary patterns that may promote disease processes or decrease the resistance of the host.

Changes brought about as a result of various forms of medical therapy are an additional general category of predisposing factors that should be considered. One of the most widely recognized dental side effects accompanying medical therapy is the marked reduction in saliva production and the changes in its chemical composition as a result of exposure of the salivary glands to ionizing radiation in the treatment of various cancers. Patients who have not shown any degree of caries activity for years may develop extensive decay as a result of the loss of the protective effects of saliva following this irradiation.[12] Marked xerostomia has also been noted in Sjögren's syndrome and the elderly, and has been related to increased caries experience.[17] Many researchers have intimated that a reduction in salivary gland function is a normal age-related change. Recent studies, however, have refuted this misconception and have concluded that the dry mouth observed or reported in the elderly is generally pharmacologically induced.[18–20] Of the 200 most commonly prescribed medications in the United States in 1980, 54 are known to cause "dry mouth."[21] Many of these medications are prescribed for the elderly.

Children are also susceptible to pharmacologic side effects. Prophylactic antiasthmatic medication has been reported to increase the incidence of xerostomia in children.[22] Feigal and Jensen have also reported that five of the seven medications most commonly prescribed for long-term therapy at a children's health center caused prolonged plaque pH depression, similar to that observed for a 10% sucrose solution.[23] The sucrose content of these medications ranged from 20 to 70% (Table 13–3). They also observed an unusual localization of the carious destruction, which corresponded to the position of medication placement in one child.

Table 13–3. Properties of Seven Liquid Medications*

Medication	Sucrose Content†	Endogenous pH	Titratable Acidity‡
Actifed	70	6.50	0.10
Pen Vee K	50	5.75	3.5
Theolair	50	5.50	0.10
Lanoxin	30	7.05	0.0
Dilantin	20	5.30	0.10
Phenobarbital	12.8	6.45	0.15
Dimetapp	0	2.70	10.0

*Source: Feigal, R.J., and Jensen, M.E.: The cariogenic potential of liquid medications: A concern for the handicapped patient. Spec. Care Dentist, 2:20–24, 1982. (Copyright by the American Dental Association. Reprinted by permission.)
†Grams of sucrose/100 ml
‡Milliliters of 0.1 NaOH to titrate 10 ml of medication to pH 7.0

Special Considerations for Special Patients

The following section outlines some of the special considerations that should be taken into account with respect to fluoride therapy for special populations.

Accessibility. Access can be a problem for both seekers and providers of dental care. This is especially true for the special populations. From the seekers standpoint, there are often physical, economic, and communication barriers that stand between them and the preventive services that the dental profession can provide. For many of these individuals, the next preventive breakthrough will not be the discovery of a better fluoride agent, but rather the elimination of the obstacles that prevent them from availing themselves of the preventive modalities that already exist. This may mean that additional resources will have to be allocated for specific populations known to be at a high risk to develop disease. Once these populations have been identified, specific caries-preventive programs can be developed.

Ability to Cooperate. A second consideration for providing services for some special patients involves gaining access to the oral cavity of an individual in need of preventive care. This may necessitate modifications of the usual routine practices with respect to positioning a disabled individual to improve visibility and stability or using devices such as floss holders and mouth props to gain additional control of movements. Another group often in need of preventive care are infants and small children. Once again, minor modifications of clinical routines can facilitate access and provide necessary stabilization and security.

It is important to assess the level of cooperation and ability each individual possesses when deciding on the appropriate preventive program. Some individuals will be able to totally care for themselves, however, modifications such as individualized toothbrush handles may be required for persons with arthritis or cerebral palsy. Other individuals will be totally dependent and the key to successful prevention in these persons will be the level of understanding and commitment provided by those responsible for their care.

Tissue Irritation. The oral soft tissues of persons who are chronically ill or who have been treated with chemotherapy or irradiation often undergo ulcerative degeneration (i.e., mucositis in a patient undergoing bone marrow transplantation). This stomatitis can become so severe that it prevents eating, drinking, and swallowing of medications. Management is usually palliative. Good oral hygiene must be maintained to minimize gingival bleeding and decrease the potential for septicemia. The use of a soft toothbrush, further softened by hot water, and dental tape have been recommended for tooth cleaning. Saline and sodium bicarbonate mouthrinses are recommended to cleanse the surfaces of the ulcers. Fluoride preparations that have an acid pH or contain alcohol or certain flavoring agents will usually not be tolerated by these individuals. Therefore, a dilute, neutral, nonirritating formulation should be considered. Our experience with patients undergoing bone marrow transplantation indicates that topical fluoride applications may have to be postponed until after the acute therapy phase, however, additional emphasis is then placed on oral hygiene measures. Upon discharge from the hospital, the patients are placed on daily fluoride gel regimens administered via customized trays.

Chronic Toxicity. In young children, the long-term ingestion of amounts of fluoride above therapeutic levels can result in enamel fluorosis. Since fluorosis can be produced only when the enamel of the teeth is developing, the chronologic period of susceptibility is relatively short. Generally, the crowns of all the permanent teeth, exclusive of third molars, have completely developed by 8 years of age. Therefore, for most individuals, age 8 is the last year when their teeth can be adversely affected by excessive levels of systemic fluoride. From a practical standpoint, however, the time of concern is even shorter. Since the problem associated with fluorosis is primarily a cosmetic one, defects would be most noticed on the incisors. Since the crowns of the permanent incisors complete development by age 4 or 5, it is felt that for these teeth susceptibility to fluorosis is not a practical problem thereafter. In other words, while excessive levels of systemic fluoride can affect the developing teeth to age 8, the period from birth to age 5, when the permanent incisors are developing, constitutes the age range when the practitioner must be especially careful that his patients do not receive systemic fluoride above the therapeutic level that can cause enamel fluorosis.

Multiple fluoride therapy is recommended in certain clinical situations so that a patient may receive both systemic fluoride supplements and topical fluoride applications. If a young child is swallowing rather than expectorating a frequently administered topical fluoride agent, such as a dentifrice, he may either have to be taken off the topical therapy until at least age 5 or the therapy may have to be modified. By age 5, the crowns of the incisors will have formed and the child may also be more amenable to following directions. Children with handicaps who may not have the intellectual capacity or physical ability to expectorate may have to have frequent topical fluoride treatments eliminated or modified until after age 5. Since tooth development and not chronologic age is the actual guide for determining the period of susceptibility to fluorosis, the practitioner should be aware that in some conditions, such as Down's syndrome, dental development is delayed. Such individuals would have a longer period of susceptibility to fluorosis.

Acute Toxicity. The acute lethal dose for a 150-pound (70 kg) man is 5 to 10 g of sodium fluoride (NaF), or approximately 2.3 to 4.5 g of fluoride (F). On a weight basis, the lethal dose for a 20-pound (9 kg) child is approximately 0.7 to 1.5 g of NaF, or 300 to 600 mg of F.[24] The maximum strength sodium fluoride tablet contains 2.2 mg of NaF, which releases 1 mg of F. A preschool child would have to consume at least 300 tablets to ingest a lethal amount, although lesser amounts would cause acute symptoms.

As a safety measure, the American Dental Association recommends that large quantities of fluoride not be stored in the home and, when prescribing dietary fluoride supplements, that no more than 264 mg of NaF (120 1 mg F tablets) be prescribed at one time. Fluoride tablets for home use are generally available in bottles of 100 or 120 in order to meet this recommendation.

Institutions that have preventive dentistry programs may stock either systemic or topical fluoride preparations in amounts that can be harmful if consumed at one time. It is imperative that these supplies be kept away from young children, patients with limited intellectual capacities, and individuals who are potentially suicidal. Fluoride supplies must be kept in a locked storage area and a current inventory list must be maintained so that it can be quickly ascertained if supplies are missing.

If an individual has ingested excessive amounts of fluoride, vomiting must be immediately induced. An emetic such as ipecac may be used or vomiting can be stimulated by placing the fingers in the back of the patient's mouth and throat. Milk of magnesia should be given to bind the fluoride in the digestive tract. In the home, copious amounts of milk may be used. Milk may not reduce the eventual amount of fluoride that is absorbed, but it will slow the absorption rate. In case of a large overdose, immediate emergency professional care, including gastric lavage, is indicated.[25]

A group that warrants special attention with respect to fluoride toxicity are individuals with chronic renal failure (CRF). Renal impairment results in diminished excretion of fluoride. Consequently, ingestion of fluoride results in prolonged elevation of serum F levels that can lead to chronic or acute toxicity, depending upon the amount ingested. We have been providing dental care for a group of children with CRF for several years. We do not recommend systemic fluorides or professional topical fluorides for this group for the previously mentioned reasons and because we have found that this group has a lower caries experience than a group of matched controls.[26] This was somewhat surprising in light of their high-carbohydrate diets, poor oral hygiene, and higher prevalence of enamel hypoplasia. We have demonstrated, however, that the plaque pH in CRF patients is elevated compared to normal controls and is related to elevated levels of urea in their saliva.[27] This protective effect may be lost following a renal transplant, depending upon the level of function of the grafted kidney.

Recommended Alternatives to Conventional Fluoride Therapy

Before discussing specific alternatives to conventional fluoride therapy, it should be emphasized that whenever possible conventional forms of fluoride therapy should be used. The basic principle, as with all patients, should be that fluoride therapy is individualized based upon the patient's caries activity, age, ability to cooperate, and the fluoride content of his or her drinking water. Based upon these factors, a program should be designed that optimizes the benefits attainable from water fluoridation (or dietary supplementation when indicated), operator-applied topical fluorides, and self-applied topical fluorides including a fluoride dentifrice. An easy-to-use, organized compilation of recommendations for conventional therapy as well as alternative recommendations for individuals with handicapping conditions that prevent them from using the conventional regimens can be found in the manual prepared by the National Foundation of Dentistry for the Handicapped.[25] The fluoride supplementation dosage schedule used throughout this publication is based upon the recommendations of the American Dental Association, the American Academy of Pedodontics, and the American Academy of Pediatrics.

An example of the conventional and alternative approaches for special populations might be illustrated by the recommendations for 3- to 13-year-old individuals with rampant caries who reside in an area with less than 0.3 ppm of F in the drinking water. A dietary supplement of 1 mg of F would be administered by means of a chewable tablet. If a child could not master the tablet, fluoride drops or an ingestible fluoride rinse could serve as an alternative. An operator-applied APF solution or gel would be used four times a year. For individuals over the age of 6 who could control their swallowing reflexes, a 0.5% APF gel would be self-applied daily for 1 month, to be followed by a daily fluoride rinse thereafter. Those unable to master the rinse would use the 0.5% APF gel applied with a toothbrush, a Toothette, or a cotton-tipped applicator. An ADA-approved fluoride-containing dentifrice would be recommended, with a toothbrush dipped in a fluoride rinse serving as an alternative.

Future Preventive Approaches for the Special Patient

Several modalities are being developed and tested that could prove to be particularly well-suited and beneficial to high-risk populations. Duraphat*, a fluoride-containing varnish, has demonstrated significant caries reductions of 30 to 50% in controlled clinical trials in Europe.[28,29] Antibacterial approaches including a caries vaccine or chlorhexidine preparation have demonstrated the ability to reduce plaque accumulation in humans.[30,31] Slow-release devices may have the potential to achieve major caries reductions by supplying fluoride for continuous bathing of the teeth.[32] Although not currently accepted for general use, these approaches may prove extremely useful in the prevention of caries and periodontal disease in the populations discussed in this chapter.

In summary, there exist special populations who are at risk to develop dental disease and who may not be benefiting from maximum preventive efforts. Emphasis should be placed on identifying and characterizing these susceptible individuals as well as determining the etiologic factors and most effective means of preventing the disease processes. They should be targeted for maximum preventive efforts using contemporary conventional methods and should receive special consideration relative to the implementation of future developments.

REFERENCES

1. Brunelle, J.A., and Carlos, J.P.: Changes in the prevalence of dental caries in U.S. school children, 1961–1980. J. Dent. Res., *61*:1346–1351, 1982.
2. Fejerskov, O., Antoft, P., and Gadegaard, E.: Decrease in caries experience in Danish children and young adults in the 1970s. J. Dent. Res., *61*:1305–1310, 1982.
3. Anderson, R.J., et al.: The reduction of dental caries prevalence in English school children. J. Dent. Res., *61*:1311–1316, 1982.
4. Koch, G.: Evidence for declining caries prevalence in Sweden. J. Dent. Res., *61*:1340–1345, 1982.
5. Marthaler, T.M.: Explanations for changing patterns of disease in the Western World. *In* Cariology Today. Edited by B. Guggenheim. International Congress, Zurich, 1983. Basel, S. Karger, 1984.
6. König, K.G.: How should prevention be achieved? *In* Cariology Today. Edited by B. Guggenheim. International Congress, Zurich, 1983. Basel, S. Karger, 1984.
7. Nowak, A.J.: Dentistry for the Handicapped Patient. St. Louis, C.V. Mosby Co., 1976.
8. Nowak, A.J.: Dental disease in handicapped persons. Spec. Care Dentist, *4*:66–69, 1984.
9. Vital and Health Statistics Series II, no. 214. Department of Health and Human Services, Pub. No. (PHS) 79–1662.
10. Rose, L.F., and Kaye, D.: Internal Medicine for Dentistry. St. Louis, C.V. Mosby Co., 1983.

*Woelm, Eschwege, West Germany

11. Legler, D.W., et al.: Immunodeficiency diseases and implications for dental treatment. J. Am. Dent. Assoc., *105*:803–808, 1982.
12. Fischman, S.L.: The patient with cancer. Dent. Clin. North. Am., *27*:235–246, 1983.
13. U.S. Bureau of Census. Projections of the population of the United States 1977–2050. Current Population Reports Series P25, No. 0774. Washington D.C., U.S. Government Printing Office, 1977.
14. Ettinger, R.L., and Beck, J.D.: The new elderly: What can the dental profession expect? Spec. Care Dentist, 2:62–69, 1982.
15. Pollick, H.F., Rice, A.J., and Echenberg, D.: Utilization of dental services by recent immigrant school children. J. Dent. Res., *63*(Abstract 1127):296, 1984.
16. Widström, E., Stenstrom, B., and Dalen, U.: Dental health of Finnish immigrants in Sweden. Swed. Dent. J., *7*:93–102, 1983.
17. Greene, C.S.: Prevention and treatment of caries in adults with xerostomia. J. Prev. Dent., *6*:215–219, 1980.
18. Baum, B.: Current research on aging and oral health. Spec. Care Dentist, *1*:105–109, 1981.
19. Chauncey, H., et al.: Parotid fluid composition in healthy aging males. Adv. Physiol. Sci., *28*:323–328, 1980.
20. Lloyd, P.M.: Xerostomia: Not a phenomenon of aging. Wis. Med. J., *82*:21–22, 1982.
21. The top 200 prescription drugs of 1980. American Drug, February, 1981, pp. 49–52.
22. Estelle, F., et al.: Ketotifen: A new drug for prophylaxis of asthma in children. Ann. Allergy, *48*:145–150, 1982.
23. Feigal, R.J., and Jensen, M.E.: The cariogenic potential of liquid medications: A concern for the handicapped patient. Spec. Care Dentist, 2:20–24, 1982.
24. Wei, S.H.Y., and Wefel, J.S.: Advances in fluoride research. *In* Pediatric Dentistry, Scientific Foundations and Clinical Practice. Edited by R.E. Stewart, et al., St. Louis, C.V. Mosby Co., 1982.
25. Fluoride Toxicity. *In* A Guide to the Use of Fluorides. Denver, National Foundation of Dentistry of the Handicapped, 1981, pp. 67–68.
26. Woodhead, J.C., et al.: Dental abnormalities in children with chronic renal failure. Pediatr. Dent., *4*:281–285, 1982.
27. Crall, J.J., Peterson, S.D., and Woodhead, J.C.: Alterations in parotid saliva and plaque in individuals with chronic renal failure. Caries Res., *18*:(Abstract 9) 156, 1984.
28. Koch, G., and Petersson, L.G.: Caries preventive effect of a fluoride-containing varnish (Duraphat®) after 1 year's study. Community Dent. Oral Epidemiol., *3*:262, 1975.
29. Murray, J.J.: Duraphat fluoride varnish: A two-year clinical trial in five-year-old children. Br. Dent. J., *143*:11, 1977.
30. Löe, H., Von der Fehr, F.R., and Schiött, C.R.: Inhibition of experimental caries by plaque prevention. The effect of chlorhexidine mouthrinses. Scand. J. Dent. Res., *80*:1, 1972.
31. Cole, M.F.: Overview and update of research on new measures for caries prevention: Food screening, immunization, and controlled release technologies. *In* Dental Caries Prevention in Public Health Programs. Edited by A.M. Horowitz and H.B. Thomas. U.S. Dept. of Health and Human Services, Pub. No. (PHS) 81–2235, 1981, pp. 148–152.
32. Mirth, D.B., et al.: Inhibition of experimental dental caries using an intraoral fluoride-releasing device. J. Am. Dent. Assoc., *107*:55–58, 1983.

Section V

PANEL DISCUSSIONS

Stephen H.Y. Wei

Moderator

We are very fortunate to have assembled today a distinguished panel of dental hygienists and dentists from all over the country including cities as far east as Boston and as far west as Honolulu. The panel members have listened, analyzed, and synthesized the research findings presented by the speakers and have come to their own conclusions as to how these ideas may be put to clinical use in their offices for the benefit of their patients. They have, no doubt, questions to raise with the speakers. But more importantly, we are looking forward to hearing their concepts of the clinical uses of fluoride in prevention-oriented practices.

Robert Boyd

As Dr. Wei mentioned, I am a periodontist and an orthodontist. My particular area of interest is the control of periodontal disease in orthodontic patients. I think most of the people here are well aware of the tremendous possibilities in that direction when looking at the average orthodontic patient's mouth. Dr. Shannon has shown that decalcification is a major potential problem for orthodontic patients, and there has been some good research on fluorides in that regard. Everybody came here today with their own special interests, but I wish we had paid more attention to the orthodontic patient. In particular, 2 years ago, Dr. Newman's study (Journal of Clinical Periodontology, 1981, by Mazza & Newman) stirred my interest in the possibilities of fluoride in the control of periodontal disease among orthodontic patients. In the last 2 years, my coworkers, Dr. Penny Leggott, and I have conducted some studies at UCSF on orthodontic patients. The first thing we wanted to do was develop a system of patient-delivered stannous fluoride application that might be applicable to the orthodontic patient, so we conducted a study on periodontal patients (Journal of Clinical Periodontology, in press) and found that a daily self-applied 0.02% stannous fluoride solution used in an irrigating system was effective in the control of periodontal disease. We now have a longitudinal study on 120 patients divided into four groups. One of the groups is using this system that I just mentioned (i.e., the daily irrigation with the 0.02% stannous fluoride solution). We have another group that is using the 0.4% stannous fluoride gel, in the method described and recommended by Dr. Shannon. A third group is using an enzyme-containing toothpaste, which according to some literature has been shown to have an effect on gingivitis, and there is also a control group using a regular dentifrice and Fluorigard mouthrinse. We felt that even the control group should have the benefit of fluoride as far as decalcification protection so the primary thrust of our work, although we're looking at caries decalcification, is to see the effect of fluoride on periodontal disease. I do not have any data on that study yet. I hope to have it at the IADR meeting next year.

Rella R. Christensen

For me, this conference has simultaneously answered questions, raised new questions, and left a number of important questions unanswered. I see that there is an imminent change coming in fluoride delivery. I am sure that all of us have noted this as we have gone through the various lectures presented here.

(1) I see changes in the types and concentrations of fluoride use. Several of the speakers have already spoken about the change from high-dose, infrequent delivery to low-dose, frequent delivery.

(2) I also see a change in the method of preparing the tooth prior to fluoride delivery. I am a dental hygienist and I've been legislated in and out of existence so many times I'm getting used to being deprived of my reason for living. I'm definitely willing to listen to possible changes, because if we no longer have to perform prophylaxis prior to fluoride application, it means that we can deliver fluoride applications more easily and inexpensively.

(3) I see a change in the frequency of use, which now places the emphasis on fluoride in the home, whereas in the past it's been the professional's responsibility to get the fluoride to the child. If we're going to have frequent fluoride delivery—twice a day or more—the emphasis is going to have to shift to the home. Our philosophy to fluoride has been shotgun, promiscuous, or however you want to call it. I guess the bottom line now is to give them fluoride every chance you get and in every form that you can deliver, hoping to find some form that they'll actually like. I can now justify this shotgun approach by what I've heard here. You may not agree with that statement, but let me tell you what I've heard here.

(1) Current caries detection methods are insensitive enough that they will not detect minimally demineralized areas in interproximal enamel. Both Wefel and Silverstone feel that these types of demineralized areas are practically ubiquitous to the interproximal area. Therefore, interproximal demineralization is controlled by frequent uses of fluorides.

(2) We heard that debridement provides the most dramatic reduction of microorganisms. (See, I've just been legislated back into existence.) But we have also heard that fluoride is necessary to cover those cases where proper and total cleansing is not possible, and anyone who has ever rendered debridement knows that those areas exist around every tooth in every patient's mouth.

(3) It was agreed that 2.2 mg sodium fluoride could be given to the pregnant mother

safely. It was also agreed—at least nobody threw any tin cans when the speaker said that it is now accepted—that fluoride passes the placenta and does deliver itself to the hard tissues of the fetus, in therapeutic amounts. We will continue to prescribe fluoride for pregnant mothers.

(4) We found that the dental integuments do not block fluoride uptake. This isn't necessarily new information, but I haven't heard it stated quite so bluntly before. I think that this is a marvelous boon to school-based fluoride programs.

How about the private practitioners? To me, it allows us now to offer alternatives. We have to make sure that we are being very honest and very candid with our patient, and our desire to line our pocketbooks has to be minimal. In the meantime, we can offer the alternatives of fluoride alone, fluoride in prophylaxis, or fluoride just delivered at home. Frankly, we have offered fluoride at no charge after prophylaxis for all our adult patients for many years. Recent research, which has taken us several years to complete, has again confirmed that a small amount of tooth surface is removed during the oral prophylaxis.

Our current regimen calls for the following:

(1) We prescribe prenatal fluoride as soon as we identify pregnancy.

(2) For the child without special problems, we use systemic fluoride commensurate with the body weight and the local fluoride concentration in water. We use a fluoride dentifrice twice a day and we recommend the use of a low-concentration OTC rinse twice a day, and we always use pit and fissure sealants. We use colored, chemically-cured sealants because we like to see what we're doing. We like the patients to be able to see it, if possible, and we definitely like the parents to see the sealants in place. There are four or five brands that we've found give approximately equal efficacy.

(3) For the child with special problems whom we would classify as the high-caries youngster, or for the youngster undergoing orthodontic treatment and other types of problems, our treatment would be the same except that we would raise the fluoride concentration. I understand from this conference that this might not be efficacious, however, we would raise the concentration to a 0.5% or 1.1% neutral sodium fluoride used as a brush-on or possibly in a preformed tray.

(4) For the adult without special problems, we recommend the use of a fluoride dentifrice twice a day, in addition to the use of a low-concentration OTC fluoride rinse twice a day, or more if they'll use it.

(5) For the adult with special problems (e.g., patients with overdentures, partial prosthesis, radiation to the head and neck, dentin sensitivity, and those with many recently placed restorations and miles of new margins that have been finished) we again recommend the higher concentration rinses in a brush-on form or with tray delivery.

I would now like to outline several important questions that I feel were not covered by this conference.

(1) How long can humans use the 0.4% stannous fluoride twice a day before it ceases to produce an inhibitory effect on the pathogenic microorganism?

(2) Which specific brand names of fluoride products deliver dependable and con-

sistent fluoride? I have listened in vain for specific brand name recommendations, and frankly, high science is fine, but as a clinician I need to know this. Neither the ADA nor our researchers would give us specific brand names. How can we know what to use and what not to use?

(3) Are combination fluoride rinses efficacious as replacements of the 4-min gel tray application of APF? This was touched on by several speakers but none would address the questions. We must have more research. Why? Because complaints about the APF gels have increased. Several panel members have already mentioned that their kids do not like the APF gels. I would like to see by a show of hands how many adults in this audience have had the gel tray treatment and kept the tray in their mouths for 4 min? Let's see your hands. That's about half the audience, and I can tell you that for those of you who haven't experienced it, for a female, it's like the first 2 months of pregnancy. For you men, it means constant nausea. Kids don't like to undergo this and even the dullest child figures out what's going to happen after about two visits to your office.

(4) On a little different level, should in vitro research replace in vivo work without first requiring parallel studies to validate the efficacy of the in vivo studies?

I would like to finish by submitting to you that there are four, not three, legs to dentistry. There are educators, researchers, clinicians, and industry providing the products that we all need, and it is my hope that all four of these legs can stand together for quality.

Jon T. Kapala

First of all, I'd like to thank Dr. Wei for inviting me to represent the American Academy of Pedodontics at this conference. I also think that the UCSF, the U.S. Public Health Service, and the ADA are to be congratulated for sponsoring this conference, and not least of all, I'd like to acknowledge again the support of private industry, for without it, the quality of this conference would be impossible.

We have been asked to comment on the clinical application of the topics discussed during the conference. In other words, do we practice what we've learned and what has been reinforced? Systemic fluorides, delivered mainly in the form of water fluoridation or properly (and I emphasize "properly") prescribed supplements are widely accepted today. Topical fluorides are beneficial in the prevention and the remineralization of carious lesions when applied in frequent low-dose concentrations, namely through the use of toothpastes and mouthrinses. Professionally and self-applied topical agents continue to benefit patients with existing caries activity. As Dr. Tinanoff mentioned today, a renewed interest in stannous fluoride preparations is as-

sociated, at least in part, with their greater potential for affecting bacterial colonization, thus reducing the potential not only for dental caries, but for periodontal
disease as well. Drs. Newbrun and Newman discussed the effects of topical fluoride
applications in the adult population when treating root caries and periodontal disease
respectively.

Clearly, fluorides in the prevention of oral disease cannot be introduced as a sole
treatment modality. Guidelines for clinical practice must be developed in order that
the entire population may benefit from our knowledge of disease processes and the
various compositions of fluoride available to clinical dentistry. Continued efforts
must be made to provide patient information regarding the efficacy of routine prevention recall. Unfortunately, patients continue to associate a recall with a tooth
cleansing service. It should be noted that representatives from the insurance industry
are conspicuously absent from this conference. Until the proceedings of this conference are made available to these agencies, coupled with their acceptance of the
importance of individualized prevention treatment modalities, it may be difficult to
achieve widespread and uniform clinical applications of the concepts introduced during this conference. Prevention cannot continue to be analogous to an examination,
prophylaxis, and a fluoride treatment.

Corrine H. Lee

I am a dental hygienist with the Oahu Head Start Program, where we have about
790 children. Our drinking water has .05 ppm fluoride and we have a high caries
rate. We have had a school fluoride supplement program for 6 years now as an
interim measure to water fluoridation, which appears to become more and more in
the distant future each year. We'll keep on plugging with our parents every year on
the need for fluoridation because it appears that those of you who are fortunate to
have fluoridation also have other needs of fluoride that would help to reduce caries
further. I would like to cut the time allotted to me so that we can hear some of the
questions answered by our speakers because I also have many questions.

Weyland Lum

When Steve Wei and Marty MacIntyre asked me to help them with this seminar, I thought it was a great idea because there is a lot of information that is very confusing and, hopefully, from this symposium, we should be able to go back to our respective areas and use all of this information in an expert fashion. I think we have accomplished this in general. So far we didn't talk about any specifics, but I believe the staff from the Council on Dental Therapeutics has clarified a lot of things and that gives me a lot more confidence in the Seal of Approval.

Dr. Bawden mentioned that we should get reprints from the manufacturers to see what they're doing. That's an excellent idea, especially if you can read a good piece of research. Unfortunately, most dentists who perhaps have not gone through the scientific process of doing research really don't understand what those reprints say. So I think it's very important that the Council on Dental Therapeutics continues to give guidance to the dental profession who, perhaps, can't properly assess the validity of the research. Another area that we should approach is how the manufacturers address the lay public. We have enough difficulty understanding things as the following case proves: 20 years ago when Gleem was being advertised, their commercial said that if you can't brush after meals, all the time, use Gleem. When I was interviewing a patient, I asked him how frequently he brushed, and he said, "I brush once every 2 weeks, but I use Gleem." I think we have to control some of those misconceptions. Fortunately, the public didn't take the Gardol shield of Colgate too seriously. I think this conference reaffirms that we really need to provide basic things such as nutritional counseling and good oral hygiene instructions, because although we talked about the availability of fluoride we don't know what's available to a particular individual. As Louis Ripa said, we have to tailor specific preventive care to the specific patient. I think it's nice that we get in to treat those newly erupted permanent molars, but in the reality of private practice, we have a problem in that our patients, especially in California are covered by insurance only up to 50 to 70%, and their policies may only cover them for prophylaxis once every 6 months. If those molars come in at different times, these services are not going to get paid. The same is true for fluoride treatment. The MediCal program in California allows for one prophylaxis a year and that's certainly not adequate for some of our patients. In summary, I'd just like to say that I think the key is to make fluoride available and to tailor the fluoride therapy to each individual patient.

Martin L. MacIntyre

Thank you for this opportunity to comment as a clinician on the findings presented at this conference. It is clear that the relationship between dental caries and fluoride is not the simple, straightforward one that most of us have been using as a basis for our clinical practice. I came to this conference with a number of questions. Many of them have been answered, and these answers will directly aid me in my clinical practice. For example, I wondered why we removed the acquired pellicle prior to giving professionally applied fluoride treatments. I also wondered why we removed plaque from areas on the tooth that were not being attacked by caries (e.g., facial and lingual surfaces). I have learned at this conference that plaque and pellicle do not appear to interfere with fluoride applications.

I have also wondered why we use the age of a child rather than their weight to determine the optimum amount of systemic supplemental fluoride. I learned that the reason was to simplify mathematics for the dentist. Since the most accurate method is by weight, I will continue to use this method for my patients to maximize caries prevention and minimize fluorosis. In this regard, all parents should be informed of the proper amount of toothpaste to use and the potential fluorosis that could occur from swallowing toothpaste. This is especially true in communities with fluoridation. Organized dentistry should work with manufacturers to create advertisements showing smaller amounts of paste on the brush.

I have not previously questioned the common understanding that a sound enamel surface, with a maximum amount of fluorapatite, produces the maximum resistance to caries. I now understand that the previously challenged enamel subsurface has a greater resistance due to the development of larger enamel crystals when fluoride is present during remineralization. Perhaps a safer, more effective, and less expensive preventive measure will result from this knowledge. I have learned that substantivity appears to be of major importance in the effectiveness of a fluoride agent, and that the stannous ion appears to have an antiplaque effect and is perhaps an anticaries agent apart from fluoride. I also learned that fluoride products, similar in appearance, are not necessarily equal due to variations in the amounts of the active fluoride ingredient, the potentially inhibiting effects of binders and abrasives, and the pH. All of these factors can affect the bioavailability of the fluoride. Since fluoride is so effective and is being used in so many different ways, it is now virtually impossible to find an ideal control group for a clinical study. The development of a highly accurate laboratory caries model or a safe in vivo model for testing new products should be a high priority.

In general, the presentations on fluoride and caries, especially the discussion on remineralization, point clearly to a chemotherapeutic treatment of small carious lesions rather than the traditional surgical approach. The discussions also suggest that the treatment is moving away from the operatory and into the home.

Answers to old questions inevitably result in new questions. Is it necessary to apply such a high concentration of fluoride (e.g., 1.23% solution) for 4 min? Is 1 min or 30 sec sufficient? Is fluoride more effective as a preventive agent after the lesion has started but before cavitation has occurred? Are combinations of fluoride agents or fluoride in combination with other agents (e.g., tin or chlorhexidine) more effective than a single agent? Is the sequencing of these combinations significant? Will newer methods of application, such as varnish, be available in the United States for use in special situations (e.g., slowly erupting molars). Could a simple and accurate method of testing systemic fluoride uptake and utilization be developed? In my judgment, development of a testing method should be a high priority.

It is clear to me that there will always be problems in clinical practice for which research has not provided definitive solutions. We, as clinicians, therefore, will have to continue to make our best decisions based on the literature and our experience, as they pertain to the individual patient.

Michael Roberts

Unlike many areas of health science where a lack of information is the source of most misunderstanding, the subjective use of fluoride may be the result of a tremendous number of studies reported under various conditions with many different fluoride compounds. It may seem ironic, but this wealth of information in a sense has made the picture less clear to the practitioner, and the therapeutic answer less obvious. Conferences such as this one can be instrumental in putting the data into focus and can also be a valuable aid to the profession. Whether you're addressing oral rinses, professional or patient appliances, and topical gels and trays or brushed-on fluorides, chewable fluoride supplements, dentifrices, or the means by which fluorides are administered to the patients, or whether you are considering neutral sodium fluoride, APF, or stannous fluoride in the various strengths available today, there appears to be a general consensus that they're all relatively effective in reducing caries. The degree of effectiveness, the preferred mode of individualized administration, and the identification of means to encourage better patient compliance to therapy are what require additional definition. The questions for future investigations include the use of fluorides prenatally and their effects on the primary teeth. Hopefully, the current NIDR-sponsored studies will shed a new light on this important area. In addition, continued research in the use of chemotherapeutic agents in periodontal disease is imperative. As Dr. Newman pointed out, although mechanical scaling and curettage are effective in changing the microbiological environment of the periodontal pocket, there are not enough dental personnel to treat all the periodontal disease that exists if we rely on this treatment modality alone. Therapeutic agents (e.g., fluorides) must be found that are effective in not only treating the type II periodontal patients who require constant care to control the disease but, more importantly, agents must be found for the type I periodontally healthy patients in order that the disease can be prevented from occurring initially. My final point, which I would stress, is that additional efforts will be forthcoming in defining preventive and palliative care for the cancer patient who experiences considerable morbidity associated with head and neck radiation and chemotherapy (i.e., mucositis or elevated rate of dental caries).

J. Keith Roberts

I am Keith Roberts from Bloomington, Indiana. I'm a pediatric dentist, not a pedodontist—there is a big difference. I've got a big advantage, I talk fast so I get 6 min into my 3. I'd like to share with you three or four ideas. I think everybody should leave here with several challenges. One of them is to go home with a new goal for dealing with future patients. My ultimate goal for my patients is a perfect mouth with a perfect bite. A perfect mouth means getting a child to age 12 or 13 without that first metal restoration in his or her mouth. A perfect bite means the teeth fit together great, look good, and work well. The key to that is basically what I call our fist, or five-sided program. Total oral hygiene control, total fluoride control, total diet control, total pit-and-fissure-sealant treatment, and early examinations and continued check-ups.

We're concentrating on one area, fluoride. I think the key to fluoride is education and I want to talk about that for 1 second. I think that we need to use the basic tenets of education (we weren't taught this in dental school): Small increments of information, immediate feedback, repetitive reinforcement with our mothers and our dads from day one. I think there's a thing called the "ladder of learning," which we need to apply to the education of patients in the uses of fluoride in our offices—it has to take people from unaware to aware and hopefully, to develop a self-interest in fluorides. If patients get involved they will start some actions that will lead to habits. In our office, we start seeing kids at 6 months old because I want to see them when their first two teeth are in their mouth. We get them back at 12 months, when they have 6 to 12 teeth and we do the first dental prophylaxis and the first topical fluoride treatment. We also repeat the treatment at 12, 18, 24, 30, and 36 months. At 36 months, when most dentists won't even see kids for the first time, we're on our fifth dental prophylaxis and our fifth topical fluoride treatment and our kids have no decay.

I'd like to leave you with three or four points:

(1) I think it's imperative that we emphasize the importance of fluoride history. It's of equal importance to a medical history. Initially, you need to take a fluoride history on your patients and you need to find out what water supply they're on, whether it is community or private, and then check the water supply, somehow, someway, find out the ppm of fluoride in that water supply. You should also find out the toothpaste they use, and don't pussyfoot around about it, just ask them what toothpaste they are using. I dare you to go back to your offices and ask everyone of your patients from this day on, what toothpaste they're using. You'd be surprised at the number of people who use Amway, Shaklee or Looney Tunes. It's amazing the number of nonfluoridated toothpastes people use.

(2) The next thing I'd like to cover is the prescribing of systemic fluorides without doing a water analysis. You should never prescribe systemic fluorides without doing a water analysis first. I just wish that somebody would sue somebody once for causing enamel fluorosis. By God, it'd be the last time someone would be prescribing fluoride in systemic form without testing the water first. I personally believe that a patient cannot get too much topical fluoride. In some circles, I would be called easy, even promiscuous in terms of my fluoride usage.

(3) In closing, I recommend that no toothpaste at all should ever enter a child's mouth during the first 12 months of life. I just don't think it's necessary. During 12 to 36 months of age, there is no need to use toothpaste during the oral hygiene procedure by the parents. Later on, a small amount, about the size of a pea, should be adequate to do a topical fluoride treatment.

In summary, we need to do initial fluoride histories and periodic fluoride updates. Never prescribe without doing a water analysis and never allow a patient to leave your office on a recall without some total home fluoride program designed for them. I think we need to get away from concepts of high concentration/low frequency, which is the way we've all been taught in schools, and get to the new concept of low concentration/high frequency. If any of you want to see what we use, let me know and we'll send you a whole package.

William R. Snaer

I've been a practicing pedodontist for about 25 years and I want to share with you my reactions to the speakers' recommendations today. Some of the recommendations confirm what I'm doing and some challenge me to do things a little bit differently. When Jim Wefel made his four recommendations, first of all determining the at-risk patients and at-risk teeth, I felt that his recommendations pretty much confirmed what I have been doing. We give fluoride treatments to about 60% of the kids who come in. But when I hear Leon Silverstone talking and see those incredible slides, I begin to feel that I'm missing the boat and that I should be more aggressive in getting more treatments to more people. Jim Wefel also talked about treating erupting teeth. You see teeth erupting all the time and so when we're seeing patient's every 6 months, I think that's probably about as close as we can get. He also talked about using pit-and-fissure sealants, which we already use. But the most important topic, however, that Jim Wefel talked about of interest to me was the carrying of the gel interproximally to the areas that we're really trying to reach, and I was really scandalized to see this model indicating that the material isn't running all the way interproximally. I'd like to ask the speakers when they return again if we can look at the scheme we have for the application of gel. It seems that we accidentally got to this business of using the trays. Maybe we need to go back to applying the gel with a swab and running it interproximally with floss and forget about using trays. Do we really need to have it on there for 4 min? Is there an approach that has a reasonable cost by which we could make these treatments more effective?

In addition, I'd like to refer to the fact that Lou Ripa mentioned that although oral prophylaxis treatments are not very useful in preventing decay, they still have some value esthetically and periodontally. It's become a bit fashionable in some quarters to look down on children's dentists who use prophy cups on children because, after all, that's not necessary to give a good fluoride treatment. It does have other values. People like to get a fresh start and their parents like to see those teeth scrubbed up, and there are a lot of other things going on at these check-ups besides just fluoride treatments—I liked that comment. He also discussed the fact that fluoride prophy pastes don't seem to be justifiable for the results, which confirms my own prejudice. When we give fluoride treatments and fluoride prophy pastes to small children, they sometimes don't feel too hot by the time they've swallowed some of the prophy paste and some of the fluoride gels. I don't think that making kids' stomachs upset is a very good way to combat the antifluoride lobbies that Leon Silverstone was talking about.

Dr. Kula talked about stressing proper fluoride supplements based upon proper knowledge of the water supply. We used to see a lot of kids in my area with enamel fluorosis. About 10 years ago, I started sending out reprints or little cards that I had made up to the pediatricians in town, telling them what the different fluoride levels in the local water supplies were. I think that was very helpful in our area and it was also positive public relations.

Furthermore, when Jim Crall was talking about the efficacy of the fluoride rinses, which we've been using for kids who have identifiable caries, I agreed with Leon Silverstone's opinion that we ought to be using these fluoride rinses more aggressively. And lastly, we've always recommended a fluoride toothpaste accepted by the ADA to our kids, but I think George Stookey's comments on selection of that kind of toothpaste might give a basis for recommending a particular toothpaste to them instead.

Stephen S. Yuen

I'd like to address my comments from my own perspective and talk about the future applications of fluoride for, not necessarily geriatric patients, but certainly adult patients. With the graying of America, we recognize that more and more of our efforts in dental care are towards the elderly, senior citizens. Even the ADA recognizes that and as all of you are probably aware, the ADA is about to launch, as soon as the House of Delegates provides the funds, a 12-million-dollar program. It is a so-called paid public education (most of us call it marketing), but if that program goes forth there'll be a tremendous output by TV and radio and press. The educational programs this time around are primarily on periodontics. They've chosen periodontics because they feel that it's primarily the adult population that determines whether or not and when the family goes to the dentist and seeks dental care. So I expect this program to be a boon to the periodontists and hygienists in the audience, and hopefully, there will be some spinoff for us general practitioners too.

Getting back to this conference, I knew when Steve called me that I was going to be out of my element; I've never been an academician. I'm certainly not a researcher and statistics bore me to death, so when I came in and started seeing those graphs and charts on the screen I knew I was in trouble. But I was very impressed as the day wore on, and I felt more and more respect and gratitude to the people in research and development and industry, and to the staff of the Council on Dental Research and the Council on Dental Therapeutics, who answered the question that a lot of people ask, "What does the ADA do for me?" Every year when we pay our dues, amounting to hundreds and hundreds of dollars, people always say "All I get is an ADA journal, and I never go to the meetings because they're always in Atlanta or some place that I can't afford to go to." But then when we start thinking about all the things that the Council on Dental Therapeutics and the Council on Dental Research have done for us, from fluorides to the high-speed turbine handpiece, then I think we ought to thank the ADA. If it sounds like I'm making a pitch, I am.

OPEN QUESTIONS AND FORUM

Wei. I'd like to thank the panelists for their insight and their contributions to this conference. We have one hour left to answer the questions that the panelists raised. I now invite all the speakers to come up to this panel and we'll address these questions.

Bawden. I have a few statements, mainly in response to Rella Christensen's comment about the products. I like to talk about products, and I opened the meeting by saying that I think there are some poor ones on the market. When we talk about topical agents used in the office and the unpleasantness of the APF system for children, I'm afraid that right now we're stuck with the situation if you want to work with something that is backed up with solid clinical trials. The simple matter of fact is that there are no clinical data on neutral sodium fluoride agents applied once every 6 months. There are no clinical data associated with the dilute rinses marketed by Omnii Gel and Gel-Kam, and in fact, the concentrations of those rinses and the manner in which they're to be applied once every 6 months almost guarantees that they're not going to be effective. Any APF product should have a pH of 4 or less to be effective, and many of the products on the market, especially the ones that taste good, have a higher pH. In fact, the other day we placed an electrode in a product on the ADA approved list and the pH was 5.2. The system is not designed to work at that pH level. You should check out the pH of any APF product that tastes really good. Take it down to your pharmacist or have someone run a pH test on it; if it's above 4, you'd better not use it, it's just that simple. As I said earlier, suspect any product that has Veegum in it. I also warn you about "laboratory data." It's a simple fact that there is no laboratory test that will predict clinical effectiveness. If plaque were the whole name of the game, amine fluoride agents would win hands down, because they're the best at removing plaque compared to any other fluoride agent, but when put in a clinical trial, amine fluorides don't work better than any other fluoride. As far as reducing caries is concerned, the extent of fluoride uptake on sound enamel, as you've heard over and over, is not necessarily a predictor. An assessment of remineralized lesions may be much more helpful, but that hasn't been used extensively as a criterion of clinical effectiveness yet. Stannous fluoride always shows less fluoride uptake at the same concentrations as other products, but it works differently: it produces a precipitate on the enamel surface and it always gets higher acid solubility reduction. That doesn't mean that its clinical performance is better or worse.

Wefel. I do not know how 4 min became established in the original literature for topical fluoride applications. In vitro studies have shown that if you cut down the application time, you get less fluoride uptake. That is not clinical caries prevention, though. The one piece of datum I know that has compared APF and sodium fluoride in terms of fluoride uptake showed that, in a much shorter time period of approximately 2 min, essentially 80% of the fluoride was taken up by enamel from APF. With sodium fluoride, however, the response was more linear and took much longer. In terms of what to do as a practitioner, the only thing I could suggest is that if the kid is giving you problems in the chair because he can't stand the fluoride in his mouth for 4 min, I would take it out after 2 min. At least you have 2 min worth of effectiveness for that patient (not that I would set out with the intention of leaving it in for only 2 min—I'd aim for the 4 min).

Participant. Dr. Newman when you're doing deep subgingival curettage and you remove some cementum, does the cementum regenerate itself?

Newman. No, the cementum does not regenerate itself.

Wei. The next question concerns the pH of topical fluoride. What products have proper fluoride concentration and pH?

Wefel. There are approximately 27 products listed in the American Dental Association's *Accepted Dental Therapeutics*. If you read that list, there are just three or four products that have different concentrations and pH. Most of the others do have acceptable concentrations and pH. Therefore, you could choose from some 20 different products that have been approved by the ADA.

Participant. Could you please comment on what instructions we should give to parents of children who've just had a 4-min topical fluoride application. I've heard different things from different people and I'm not sure what's current as of today.

Ripa. The admonition of "don't eat or rinse for half an hour," was somewhat empirically derived, as I believe Dr. Stamm indicated. There is, however, some rationale for doing that, and I would certainly continue to say that after you've had your topical fluoride, do not eat anything and do not rinse for the next half hour, but you should spit.

Wei. The stability of stannous fluoride products has been discussed. On the table showing A, B, C, D, E, F brands by Dr. Tinanoff, which brands are more stable? Would Dr. Tinanoff like to give us an answer?

Tinanoff. Brand F is the most stable. Brand F is Gel-Kam.

Wei. What does "stable" mean? Does everybody follow that or not?

Tinanoff. OK. It has the correct pH, whereas some of the others do not. It has 100% of the fluoride that should be there. It has 100% of tin that's supposed to be there. Brands that do not have the correct pH or do not have the correct amount of tin don't have complete antiplaque effects. Gel-Kam has the perfect pH and the perfect tin concentration. It's important to have brands that have, like Jim Bawden said, the perfect pH and the perfect fluoride level.

Wei. The next question is among the ones submitted for the Council on Dental Therapeutics. Dr. Naleway, there seems to be a lot of differences in the stability of different stannous fluoride products and yet they're all accepted by the ADA. How does the Council justify such positions?

Naleway. First of all, the Council's position has been based solely on the clinical results. The Council's acceptance has been based solely on proving the product to be equivalent to clinically tested results. In terms of the stannous ion content, all the manufacturers have been working with us to improve their formulations and, we believe, all have made serious attempts to improve the stability such that the majority of products now are in the 80% or above range.

Wei. The other question is: If, in fact, a manufacturer sells a product without coming to the Council, what is the position of the Council in terms of compliance or monitoring?

Naleway. The ADA's program of acceptance is a totally voluntary procedure by which the manufacturer comes to us for our review. Products that do not come to us for review would not be accepted and would never be reviewed.

Participant. I found very early on when I was prescribing fluoride trays and fluorides for patients that they would often come back and say "I can't use it." So what I try to do now is give them three or four sample bottles of different fluorides in the hope that one of the solutions they select will be beneficial. I also tell them that if they tell me which one they liked or had the least side effects from at their next visit, I would provide them with that particular product. I've been pretty successful with this method. I tell the patients to put the fluoride tray in their mouth while they're taking a shower. If you take a normal 5-min shower, you can take the trays out of your mouth just before you're ready to rinse off and rinse them off and then you don't have to worry about swallowing the fluoride. I have trays for myself that I use periodically. I hate to stand over the sink for 4 min, whereas it is very easy to do it in a shower.

K. Roberts. Dr. Mitchell, I would like to see the audience give you a vote of support. I am extremely concerned about future dental conventions and dental meetings where Joe Blow dentist from Anytown, USA comes in and sees a product at the ADA Convention Exhibit Hall. A lot of people feel strongly that if something is exhibited there, it must be there because it's OK; otherwise, it wouldn't be there. I think you need to bring our feelings and concerns about this to the attention of the Council on Annual Sessions. I was talking to Dr. Naleway's associates and learned that the new rules and restrictions on advertising have changed. There are also ads in the ADA journal for products that are not accepted or accredited because the ADA wants the income, and I hate to see this happen to our meetings. I think you ought to relay to Chicago how strongly we feel about the exhibition of products that have not been approved.

Wei. Dr. Silverstone, if a white spot lesion could remineralize as much as 30 microns deep, does this mean that gross caries cannot be remineralized?

Silverstone. If by gross caries you mean a lesion that's already cavitated, then the answer is it cannot remineralize in the way I showed today. Even within the cavity, things can happen, however, but you cannot achieve the sort of results shown on the slides if the surface is truly cavitated. When you do a topical fluoride application, you're treating sound enamel, and if fortuitously you manage to treat those sites with small lesions, then you're going to have the benefit anyway.

Horowitz. It's obviously true, you can't remineralize a lesion where the surface enamel is gone, but one can arrest an overt lesion. Therefore, I think fluoride also has a role in stopping the progression of decay towards the pulp. So, while it may be too late to end up with a better tooth than you had before you started, I certainly think that fluoride can stop or slow down the process.

Silverstone. That is absolutely true, and the problem in remineralizing enamel is trying to remineralize it through the intact surface. If the surface is open you can achieve a great deal. In terms of remineralizing dentin, thermodynamically it's almost child's play compared with enamel.

Wei. Please comment on the relative relationship between arthritis and fluoride, as two to three patients have mentioned fluoride was bad for them.

Tinanoff.   I don't think there's any evidence either way. I've never heard of that before.

Wei.   Do you recommend that any cancer patient undergoing radiation therapy should undergo immediate fluoride treatment, topical and/or systemic? I think that answer is quite obvious. Everyone is nodding their heads yes.

Wei.   The next question is addressed to Dr. Naleway or Dr. Kula. What is the *clinical* effect of fluoride on porcelain or composite restorations?

Kula.   Dr. Van Thompson and other researchers have shown that the 1.23% APF solutions and/or gels have some surface effects on porcelain. When these materials are examined in a scanning electron microscope, it appears that there's a roughening effect on the porcelain. Dr. Thompson tried this in vivo on a patient, and after one application of the 1.23% APF gel, he saw visual changes in the porcelain. Dr. Thompson and I have been working on composite resins and we have completed some in vitro studies. Our first study was with three different composite resins. One was a microfil with silicon, another was a conventional composite with quartz filler particles, and the third was a strontium glass-filled material. All three materials that were treated experimentally with five 4-min exposures to APF gel had significant weight loss compared to the control. The strontium glass-filled composite lost significantly more weight than the quartz-filled composite, which lost significantly more weight than the silicon microfil composite. Since that time we've tried that on filled and unfilled resins, or sealants. We have seen visual changes and changes in the scanning electron microscope on the filled sealants but not the unfilled sealants. The changes seem to be occurring with the filler particles, although the data are not complete yet because we're just documenting visual changes so far. We've started work on about 15 to 20 commercially available products. We have compared visual changes in surface reflectivity in these materials when they were treated experimentally and they have a definite surface roughening effect.

Wei.   What is the clinical effect?

Kula.   There has only been one in vivo study and that was on just one patient. It seems to have a real effect. Clinicians are reporting to us the effects on porcelain when they give their patients 0.5% APF gel to use on a daily basis. We have seen no effect in in vitro studies using neutral sodium fluorides, though.

MacIntyre.   This question is for Dr. Silverstone. It appears that economics, ethics, and the failure to find control groups make it very difficult to do future clinical studies. The only hope is to have a model with plaque, pellicle, and bacteria on extracted teeth and to come as close as we can to the real clinical situation. How far away are we?

Silverstone.   Like I say, how long is a bit of string? I think we're getting closer. In vitro, you can only achieve so much no matter how sophisticated your model is. The next step is, and we've already started this, to do some in vivo work using the in vitro technique. For example, human teeth containing standard lesions are being put in appliances and we can, therefore, monitor these both before and during experimental regimes. These are very small numbers naturally, by the nature of the design, but at least it'll take the in vitro step one step closer to where we are in terms of a clinical approach. To have a full blown clinical trial, as we know, costs a lot of money and it takes time.

MacIntyre.   Would you be comfortable once you do that with recommending a new product or technique?

Silverstone.   Well, everything's a series of steps. You go that way and then you say "the next step is going to be a pilot clinical trial" and a pilot has to be designed and carried out. It takes time, of course, but you have to go through a logical series; otherwise, misleading results of one step in the series could have an effect on the next step. So you have to go through the thing logically.

Schiff.   We have heard about the successful treatment of periodontal pockets with stannous fluoride rinse. You forgot to mention how often you had done that before the success came about?

Newman.   This was done over a period during the last 3 years. It was done in the original studies with Mazza, in which 10 to 12 patients were used. As I mentioned yesterday, what we wanted to do first in those studies was to determine efficacy. Can we kill bacteria in the pocket, and how long does this effect last? The next step is to try it in humans, such as Dr. Perry's study, and by the design of his study, there was very little difference between patients who had scaling in a deep pocket and those who had fluoride treatment. One application was used.

Dr. Horowitz asked about additional studies that might show the effect of stannous fluoride over and above that which occurs with regular toothbrushing. There are several studies. One study reported in the Journal of Clinical Periodontology and the Mazza study presented at the last IADR meeting both suggest that there is an additive effect with using stannous fluoride. I think it's important to state that it was a safe approach, and efficacy needs to be monitored and developed. Conceptually, it's a terrific idea to get that relatively low-concentration fluoride applied on a daily basis to exposed root surfaces. It results in decreased plaque, bleeding, and motile bacteria population.

Participant.   What is the product's name that you used in the periodontal pockets? Gel-Kam is kind of viscous and it's tough to get it down into deep pockets with the syringes that you talked about.

Newman.   We use the product made by Scherer Laboratories. We're pretty comfortable with that. We use a nonviscous, 1.64% stannous fluoride solution and flush the pockets.

Participant.   The other question that I had is right in line with this. I guess we saw the time-release fluorides that were bonded on the tooth surface. Has anyone done research with time-release fluorides that could be placed into deep periodontal pockets in areas that are unmaintainable?

Newman.   Not that I know of. The only time-release products I know of have been those used with hollow-fiber devices using tetracycline by Goodson's group in Boston.

Tinanoff.   I did a clinical trial in Switzerland, because it is not allowed in the United States. We studied slow-release stannous fluoride placed intercoronally, mixing stannous fluoride with a polycarboxylate cement. Because I was using Class II lesions, these devices were in the periodontal pockets. The high concentration of stannous fluoride in the slow release system was harmful to the gingival tissues. It caused gingival irritation, so a high concentration device left there a long time is probably not a good thing.

Wei.   I also would like to ask Dr. Newman and Dr. Tinanoff to qualify the status of the use of stannous fluoride products or any other fluoride products on gingival tissues or in periodontal pockets. Is this experimental research work, or is this accepted current practice, because we don't want everybody to start doing this without carefully selecting the patients or defining specific indications and contraindications.

Tinanoff. The slow-release systems are definitely experimental, but if you think that the Keyes technique is an accepted therapeutic regimen now, then I don't see anything different with flushing pockets with stannous fluoride. Stannous fluoride is safe, it's been used for years. I myself think it's experimental, but certainly practicing dentists are using things that other researchers also consider experimental.

Newman. I think we want to divide up the flushing of the pockets, which I think is an experimental technique. I also happen to think it is a safe technique, and we're going to continue to develop information about it. The use of the daily 0.4% stannous fluoride gel is essentially past the experimental stage to the point of the clinical trial. This is the concentration that has been in dentifrices that have been used for years and years, and I think conceptually are definitely worth pursuing in the periodontal patient.

Mitchell. I would just like to make a statement about the stannous fluoride gels relative to the FDA and the Council. The FDA recognizes the products as an adjunctive for an anticaries benefit. It does not recognize them for an antiplaque effect nor for reducing sensitivity of dentin or any other therapeutic modality. This is one of the reasons that the primary feature that we looked at in these products is available fluoride, because that's what produces the anticaries effect. The Council has accepted these products and has requested that labeling be limited to those areas that have been clearly demonstrated. These products are for rampant caries and special patients who need that as an adjunctive treatment in addition to a fluoride dentifrice, not in place of a fluoride dentifrice. At the present time, that's the information that manufacturers have submitted to us and that's the basis upon which they've been accepted. That does not mean that there is anything wrong with using them the way that was presented here today but if questioned or asked, that's the official response that you'll find the FDA and the ADA giving for these groups of products.

Wei. Fluoride products are approved for anticaries applications but are they available for prescription by dentists for nonapproved uses as well? Are the speakers encouraging practicing dentists to buy fluorides for uses not approved by the FDA? Are the speakers involved in or aware of new drug applications for FDA approval of fluorides for these new indications (by that they mean hypersensitivity, antiplaque activity, periodontal disease, etc.)? Would you like to address that question?

Mitchell. Yes. It's the position of the FDA that a practitioner may use any approved drug for a nonapproved use. The responsibility for that, of course, is solely in the hands of the practitioner. In this case, efficacy is the primary issue and safety is not an issue relative to the use of these products.

Participant. A few minutes ago the panelists mentioned that it would be wise for us as general practitioners to ask our patients which toothpaste they're using and then to recommend a specific toothpaste. How do we know which specific toothpaste is best?

Stookey. I think, that as a practitioner, you probably have on your shelf your first and foremost guide (Accepted Dental Therapeutics) to the selection of all types of products, including fluoride dentifrices, topical fluorides, or whatever you want to call them. I think in that listing are all the approved products that were cited in Dr. Naleway's presentation. That's a handbook that you should use continuously when a salesman comes to your office with whatever product; your first real clue to whether or not he's selling you something that is accepted, useful, and effective should be the Accepted Dental Therapeutics' listing. This is a very inexpensive book and it's reupdated every 3 years. That should be your best guide.

Head Start Participant. I'm from Santa Clara County and I'd like to add to what Rella Christensen mentioned as a "four-legged stool for prevention"—I believe it's a five-legged stool. We always forget the patient and the consumer. After working with 60,000 children in a fluoride mouthrinse program, you become half consumer and half provider. For years, in our county we've worked with the children and their parents and teachers as coworkers. I'd like to ask a question of Dr. Carlos, and it may seem to be a simple question. What is the relationship between brushing and the absorption of sodium fluoride rinse? We have quite a controversy going: some of our doctors say "unless you brush thoroughly and professionally clean the teeth, there will be no absorption from your sodium fluoride rinse," and there are documents that say both, and I am really confused. Would you clear up my doubt?

Carlos. I think so. Actually, I think Dr. Ripa should answer this because he has reported on the result of some studies of the very point of whether it is essential to clean the tooth by a professional prophylaxis or brushing by the children. The answer seems to be it's not. You don't have to do that. You can be confident you're going to get good anticaries results.

Mellberg. I want to thank George Stookey for a good review on the fluoride dentifrice situation. I also want to make a few comments on some of his conclusions. I think he concluded that perhaps the sodium fluoride toothpaste was really more effective than a sodium monofluorophosphate toothpaste and he based that, not on clinical trials, because he admitted that there were no head-to-head comparisons of these current formulations, but on conclusions based on laboratory experiments. Since I did some of the laboratory experiments that he referred to in addition to other ones that I think are pertinent over the last 20 years or so, I would like to say we have to be very careful about how we interpret laboratory experiments. For example, a number of years ago, in collaboration with NIDR and other investigators, we analyzed the fluoride content of teeth that had been treated with neutral sodium fluoride and APF preparations within the same study and at the same fluoride concentrations and frequency, and we found that, at least under frequent applications, the APF formulations put a considerable amount of fluoride into tooth enamel, whereas the neutral preparations put in only a modest amount.

In one study we used gels containing 5,000 ppm fluoride or 0.5% fluoride. In another study we used fluoride rinses with 200 ppm, and in both of these studies there was no difference between the caries inhibition due to the neutral sodium fluoride and APF rinses. Therefore, fluoride uptake in these cases wasn't really pertinent. I should admit that this was sound enamel and I'm going to be one of those people who say that we shouldn't be looking at sound enamel, but we should be looking at white spot lesions. One of the studies that George Stookey referred to was a white spot enamel study that we did in vivo. He stated that the sodium fluoride preparation gave better fluoride uptake than the sodium monofluorophosphate dentifrice, which numerically was correct, but statistically it was not significant. The numbers he had in his table were a little bit misleading, because you'd have to see the shapes of the curves. The first point on the curve looked as if there was a big difference, but if you integrated the area under the curves, you'd find that there was really little or no evidence for saying that one dentifrice is more effective or less effective than the other.

Stookey. I'd like to clarify a few points. Jim, I did not base my decision upon laboratory data. I put together a review of every clinical study I could find in the literature, and I tried to look at those objectively and answer the question that Dr. Wei had asked me. The first five listings for that decision were based entirely on clinical data. I used the laboratory data to help understand how that could be, and that was the only use of the laboratory data. It was explanatory data; I did not make the decision based upon laboratory data. It's pretty obvious that we can never analyze individual products, so we have to deal with those kinds of analysis on a system basis, which is basically what I did. I tried to look at the fluoride systems rather than individual products or compositions.

Participant. This question, I believe, is for Dr. Tinanoff. If I understand correctly, the ADA has approved the stannous fluoride at 80% stability and above. What I want to know is, in plaque inhibition studies, will the 80% give you the same results as those that have higher tin availability or better stability data?

Tinanoff. In my presentation, Brands B, C, D, and E all showed that they did not reduce bacteria on the wire as well as Brand F. Therefore, I guess you have to draw your own conclusions. I would also like to pass that information on to the ADA Council on Dental Therapeutics.

Participant. We don't have an effective tracking system for migrant children and their fluoride history, and you recommend that preschool children not rinse with fluoride (although I've done some pilot studies myself and found that if preschoolers are taught to rinse and practice rinsing, they do not swallow it). I would like to know what you recommend as the most beneficial, most cost-effective regimen for fluoride supplement in Head Start Programs?

Ripa. Actually, Dr. Wei wrote a Head Start manual on just that topic. In terms of the preschool child in a Head Start program, however, a daily fluoride tablet program of appropriate dosage is probably the least costly and most effective. Also, we now have the Head Start children brush with a fluoridated toothpaste daily, but that should be well supervised and controlled. In regard to your statement about preschool children being able to rinse, I definitely agree with you. They can. There have been statements made in the dental literature, however, that the swallowing reflex is immature before age 5. Well, children have a swallowing reflex before they're born and within the first couple of years after birth, all of the anatomical structures necessary for a proper swallowing reflex are present and in operation, so the Head Start kids clearly have swallowing reflex and it's a simple question of teaching them to swallow or not to swallow. In terms of an answer to your question, I'd go with the tablet.

Same Participant. And for migrant children, even though they're coming from various areas and the fluoride histories are unknown, do you still recommend the same?

Ripa. I think you could only treat them as long as you have them.

Participant. I'd like to just finish off that topic because I work with Head Start children also. These kids are in programs as short as 9 or 10 weeks so we treat them as long as they're there. If you try to monitor where these families go, it's a virtual nightmare. You cannot track fluoride so the only thing left is some kind of topical fluoride. Head Start standards require that these programs provide professionally applied topical fluoride, so they end up burning a lot of money on that procedure and actually run out of treatment money. One of the possibilities is to recognize the use of fluoridated toothpastes twice a day as the best alternative because it at least reinforces an important habit and they get some fluoride benefit. Would this panel advocate accepting that kind of a regimen? Or, is the only alternative professionally applied topical fluoride?

Carlos. Well, I don't know. I don't think anyone has done any studies of children under the conditions that you have to deal with; that is, where you have them for a brief period of time and then they're at risk without preventive or supervised preventive therapy for a long time. Sure, I would certainly try my best to get them to use the fluoridated dentifrice first, and if you can, I'd try to supply them with that and urge that they keep on using it. Whether they will or not I suppose you have no way of knowing. Would I give them a professionally applied fluoride treatment of high concentration gel? Sure, if I could afford it. I mean, this is a situation in which you have no really good guidelines and you're just going to do the best you can in the time available to you and hope that they follow through, perhaps, with the dentifrice. I haven't worked with migrant children but that would be my approach.

Silverstone. I agree with what Jim said.

Wei. How long can we use 0.4% stannous fluoride gel twice daily before it ceases to be effective? That is one of the questions that Rella Christensen asked.

Tinanoff. We don't know yet.

Wei. So it looks like a study is definitely needed in this area.

Horowitz. I'd like to make a follow-up comment on the children of migrant workers in Head Start programs. It's true you can't track those children to where they're going but you know where they are now, and if they are not in an optimally fluoridated area or getting drinking water with adequate fluoride, you can give dietary fluoride supplements during the period that they're there. If they're used properly and chewed and swished before swallowing, you're delivering a topical effect daily from the use of the fluoride plus the systemic value that it may have during the period that they're in the area. I wouldn't give them an added supply to take with them because you don't know where they're going, but while they're there you could give it to them.

I can't say I was fully convinced by Dr. Newman in terms of the value of 0.4% stannous fluoride, although you did call for additional research, but I'm confused again after hearing Ed Mitchell's comment on the basis for the acceptance of 0.4% stannous fluoride. I thought I heard you say that the Council recognized it on the basis of its anticaries effect when used in conjunction with a fluoride dentifrice. What is its recommended method of use? I still don't know what research exists to indicate that the dual use of the two products, which both contain 1,000 ppm of fluoride, in succession produces an additive benefit?

Naleway. Studies were done by Dr. Shannon in the VA group and one was done on orthodontic patients to decrease the demineralization potential. The studies were done in these populations because they were undergoing aggressive oral hygiene practices, which included the use of these gels on a daily basis.

Ripa.   I think part of what we're answering or what we're discussing goes back to Mrs. Christensen's questions, and if I may paraphrase them, they were:

(1) What products are recommended to be used?
(2) When or how should they be used?
(3) What products don't you recommend being used?

In terms of the products that are recommended to be used, clearly you should go through the list of Accepted Dental Therapeutics, which is published in November of every year. In terms of when or how to use the different fluoride agents (and by that I mean fluoride dentifrices, operator-applied topical fluorides, self-applied fluorides, and systemic fluoride supplements), the manual published by the National Foundation for the Handicapped answers that question not only in terms of when to use these products individually, but more importantly, how to use all of them at once, appropriately, on any individual patient. So really, the answers to those two questions are basically already written down and available to everybody who takes the time to find the material. In terms of what products not to use, I think that is a problem, because there is no unanimity of agreement, particularly in two areas. Let me speak only in terms of caries prevention, because I don't know a hill of beans about preventing periodontal disease other than I think if I brush my teeth correctly and floss, I'm in pretty good shape. At any rate, in terms of caries, there are two procedures presently being used and recommended by several companies that many knowledgeable people in the area disagree on. One of these procedures is the use of a multiple fluoride mouthrinse in the dental office, using the APF and stannous fluoride rinses. There's absolutely no clinical evidence to support that this particular procedure works. What is even worse about recommending this particular procedure is that it is being recommended in place of something that we know does work. So I, for one, certainly cannot recommend that particular procedure. Now the procedure that I think engenders the most controversy is the use of the 0.4% stannous fluoride gel, and here I think we are, in part, talking about definitions. For me, the definition of a proven cariostatic agent is clearly different from the definition that Dr. Mitchell just gave. For me, a product, technique, or agent is proven to be effective in terms of reducing dental decay when it has been proven in repetitive, controlled, independent, long-term clinical trials, and the 0.4% stannous fluoride gel does not fill that definition. Additionally, as Dr. Horowitz indicated, most patients, especially children, are already brushing with 1,000 ppm fluoride. They're doing this with the dentifrice that they use. So that brushing once or twice a day already with a 1,000 ppm fluoride dentifrice and then brushing one more time with a 1,000 ppm fluoride gel and you're going to get these terrific benefits doesn't make sense to me. If it does make sense, why don't you simply recommend brushing their teeth two or three times a day with the dentifrice that they're already using? I think this is an area where there is a controversy.

Wei.   I'd like to thank the speakers and participants for an outstanding session of stimulating discussions. The conference is now adjourned.

INDEX

Page numbers in *italics* refer to illustrations; page numbers followed by a t refer to tables.